DATE DUE

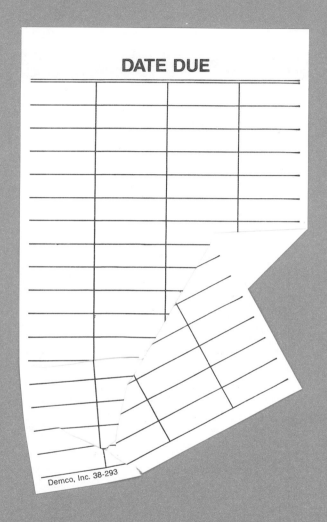

Demco, Inc. 38-293

Canada's National Aviation Museum

ITS HISTORY AND
COLLECTIONS

Canada's National Aviation Museum

ITS HISTORY AND COLLECTIONS

Kenneth M. Molson

WITH SPECIAL SECTIONS BY

R. W. Bradford and P. A. Hartman

National Aviation Museum

National Museum
of Science and Technology

OTTAWA 1988

PRINTED AND BOUND IN CANADA

Canadian Cataloguing in Publication Data

Molson, K. M.
Canada's National Aviation Museum:
its history and collections

Issued also in French under title:
Le Musée national de l'aviation du Canada.
Includes index.

ISBN 0-660-12001-1

1. National Aviation Museum (Canada) –
History.   2. National Aviation Museum
(Canada)   I. National Aviation Museum
(Canada).   II. Title.

TL506 C3 M64 1988
629.13′074′011384
C88-099107-0

ILLUSTRATIONS

PAGE 1
Col. W.G. Barker, VC, warming up the
war trophy Fokker D.VII which he
flew in the New York – Toronto Air
Race. This is in front of a hangar at
Leaside Aerodrome; another D.VII
can just be seen in the hangar.
(CWM/A.G. McLERIE COLLECTION)

TITLE PAGE
The Sopwith Snipe (left flown by
W/C D.P. Wightman and the Sopwith
Camel flown by W/C P.A. Hartman
in formation over Rockcliffe, in June
1967, both airborne for the first
time in almost 50 years.
(DND PC-67-454)

PAGE 11
Aircraft in the new Museum in their
protective coverings.
(MALAK)

PAGE 95
Last flight!
After the Bellanca CH-300
Pacemaker arrived at Rockcliffe on
May 29, 1964, from Juneau, Alaska,
newsmen asked pilot Gath Edward
if he would make a brief flight for
photos. It is believed to be the last flight
of any member of the long-distance
record-breaking Bellanca family.
(KMM)

Contents

Foreword

The invention of the practical flying machine at the beginning of the 20th century was to have an enormous global effect on human affairs in peace and war. Here in Canada the flight of the *Silver Dart* at Baddeck, Nova Scotia, on February 23, 1909, just five years and two months after the Wright Brothers' success, was to trigger a series of events that would establish Canada as one of the leading aviation countries in the world. During the Great War of 1914-1918 Canadians demonstrated their ability to manufacture aircraft in large numbers and to fly and maintain them. Indeed, in that first great struggle in the air, Canada's contribution was far out of proportion to the size of its population, and the resulting reservoir of highly skilled people returning in peacetime was a significant factor in the growth of aviation in this country in the post-World War I years.

The commercial use of aircraft in Canada began as early as the summer of 1919, primarily in the development of our natural resources and in meeting the transportation needs of this enormous country. Some measure of what was achieved in the early years can be seen in the aviation statistics of 1934, which recorded that Canada carried more air freight in that year than any nation in the world.

During the twenties and thirties the Royal Canadian Air Force was carrying out work that was essentially civilian in nature and in so doing became quite remarkable in world history. It was not uncommon to see two bush aircraft of the same type land in a northern lake, one with civilian markings and the other bearing the military roundels of the RCAF. That period saw the development of the flying clubs that proved to be so vital as a resource for training aircrew in World War II. It was in those same years that original Canadian aircraft designs appeared of aircraft that were particularly suited to the rigorous demands of this country, and the first regularly scheduled airmail and passenger services in Canada began.

The storm clouds of the late 1930s burst into global conflict, and again Canadians were asked for a very special effort. They responded by manufacturing over 16,000 operational and training aircraft and by training thousands of air and ground crews from Canada and the British Commonwealth. The training demands required the building of airfields all across Canada, many of which are still in use today. The immediate postwar years saw the development of specialized Canadian-designed and -built aircraft for both civil and military use, such as the de Havilland Canada family of high-performance STOL aircraft, which are known around the world, and the Canadair Argus that served so long and well in the protection of the sovereignty of this country.

Our capability in the realm of jet flight was established with the flight of the Avro Canada Jetliner in 1949. This was followed in 1950 by the Avro Canada CF-100, a dependable, long-range, all-weather interceptor, and the spectacular, supersonic Avro Arrow, which first flew in 1948. Airline operations sprang up across the country and expanded into the great international services we know today. Bush and arctic flying reached a new intensity, and general aviation was developing so rapidly that it became more and more visible to the public. Indeed, it is a fact that per capita more Canadians own or fly aircraft than any other people in the world.

The preservation of this special Canadian aviation heritage is important to us all so that we can understand our place in this remarkable age of flight. Because of the dedicated farsightedness of a number of individuals over the past seventy years, and especially the efforts of a group headed by John Parkin almost fifty years ago, Canadians can be proud of their National Aviation Museum, which houses a major aeronautical collection, one that attempts to show Canadians our aviation heritage within the context of world development.

The story of how the National Aviation Museum came to be is one that will interest all Canadians. How appropriate that this important work has been undertaken by the founding curator of the original National Aviation Museum at Uplands, Kenneth M. Molson. Ken Molson's lifelong involvement in aviation and his commitment to the preservation of aviation history in Canada, even long after his retirement from the National Museums in 1967, has established him as the leading aviation historian in this country, as well as bringing him international recognition. His book traces the origins of the NAM collection to the present time, while telling of the people and decisions that brought into being one of the world's most significant aviation museums.

R. W. BRADFORD
Associate Director
National Aviation Museum

Acknowledgements

In the preparation of a history the author is usually indebted to many for assistance in providing information and photographs, and this book is no exception.

First, the preparation of this book would not have been possible without the complete co-operation of all the staff of the National Aviation Museum, which has been willingly given and is most appreciated. R.W. Bradford, Associate Director, and A.J. (Fred) Shortt, Curator, have been more than helpful, as they have reviewed the complete draft of the book and contributed useful comments. Bob Bradford has kindly contributed the foreword and the section dealing with the new Museum building, which is especially appropriate, as he has been closely associated with it since its earliest inception.

L.F. Murray, former Secretary and Associate Director, Canadian War Museum, and H.A. Halliday, Assistant Curator of War Art, Canadian War Museum, have been helpful in supplying information and photographs. R.V. Manning (W/C, RCAF, retired) has kindly provided information relating both to his years with the Canadian War Museum and to his period as RCAF Air Historian, and he also reviewed related portions of the draft of the book. P. de L. Markham (W/C, RCAF, retired) contributed his helpful recollections of events leading to the Museum's acquisition of the Rockcliffe site.

I am grateful to P.A. Hartman (W/C, RCAF, retired) for contributing his informative account covering the flight characteristics of the Museum's aircraft, which has added appreciably to the Museum's story.

In the preparation of the pre-1960 history of the Museum, the late Dr. J.J. Green was very helpful by initially keeping me heading in the right direction and then reviewing and commenting on that portion of the draft. Also, I am grateful for the kind assistance of Dr. A.W. Tichnor, Senior Archival Officer, National Research Council, in making available the records of the pre-1960 period which formed the basis of this account of that time.

8

I must record my appreciation of the fine cooperation of John A. Griffin and John F. McNulty, who have contributed much in recording our aviation history. Both responded, as always, quickly and helpfully to requests for assistance in their respective fields of RCAF aircraft history and aviation photographs.

Under the direction of Wendy McPeake, the Publishing Division has produced a fine publication. Adele Lessard carried out the French version with enthusiastic efficiency. I am indebted to Robin Brass who edited the text with his usual expertise and to Frank Newfeld for his exceptional design. Finally, my gratitude is expressed to all of those listed below for their assistance in supplying information and/or photographs: P.M. Bowers; B. Buchanan, Air Canada; D.W. Carter; Dr. W.A.B. Douglas, Director of History, DND; the late I. Geddes, Canadair Ltd.; G.G. Holmes; F.W. Hotson; P. Jarrett; the late A.N. Lecheminant; H.H. Lippincott, United Technologies Corp.; M. Lundy, DND; R.C. Mikesh, L. Milberry, R. Morris, National Air and Space Museum; P. Robinson, Public Archives Canada; the late A.G. Sims and Mrs. Sims; J.W. Underwood; H. Whittington, *Canadian Aviation*; H. Wynne; C. Vincent, Public Archives Canada.

EDITORIAL NOTE
Aircraft specifications are given in metric and imperial measure. The original manufacturer's measurements are followed, in parentheses, by the converted units. All specifications are given for aircraft as land planes, with the exception of the flying boats.

Creating a National Aviation Museum

W.R. Turnbull with his successful variable-pitch propeller at Camp Borden, June 1927.

(W.R. TURNBULL/NRC)

To start a history of the National Aviation Museum with its opening in 1960 would do an injustice to earlier efforts to establish a museum and collect specimens beginning about 1930. So the story must start with some of Canada's aviation pioneers and some farsighted early contributors. Appropriately, the first person to contribute to the collection, albeit indirectly, was Dr. Alexander Graham Bell. Dr. Bell, a Scot by birth, came to Brantford, Ontario, in 1870, shortly settled in Boston, and later moved to Washington, D.C. In 1883 he established a fine summer home near Baddeck, Nova Scotia. While Dr. Bell is primarily remembered for the invention of the telephone, he became interested in aeronautics as early as 1877. It was S.P. Langley's experiments, however, that led him to take up experimenting in 1891.

With the success of the Wright brothers to inspire him, Bell suggested the formation of the Aerial Experiment Association in 1907 to design and build flying machines; it was funded by Mrs. Bell. After the A.E.A. had built four successful aircraft at Hammondsport, New York, Dr. Bell, with his strong sense of history, suggested that the fourth A.E.A. machine, the *Silver Dart*, be brought to Baddeck. There, on February 23, 1909, it made the first powered heavier-than-air flight in Canada.

Dr. Bell then suggested the creation of Canada's first aviation company, the Canadian Aerodrome Co., and assisted F.W. (Casey) Baldwin and J.A.D. McCurdy in its formation. Dr. Bell preserved relics of his experiments, the A.E.A. and the Canadian Aerodrome Co. at his Baddeck estate and, as will be seen, some of them were given to the Museum; others led to the formation of the Bell Museum at Baddeck in 1956.

Another pioneering Canadian, Wallace Rupert Turnbull, from Saint John, New Brunswick, graduated from Cornell University in 1893 and, after doing graduate work at Cornell and Heidelberg, Germany, went

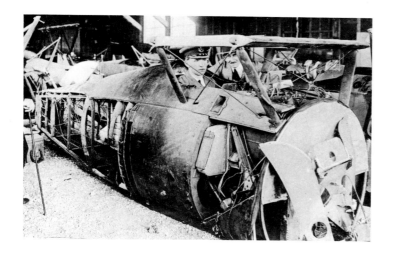

on to become Canada's first aeronautical scientist/engineer. In 1901 he set up a consulting practice at Rothesay, New Brunswick, where he built Canada's first wind tunnel in 1902, possibly the eighth in the world. Turnbull was one of a handful of engineers in those early days who really understood the problem of flight; the majority of early experimenters were lacking in technical background. Turnbull specialized in propeller design and developed a successful electrically operated variable-pitch propeller, which was tested in 1927. He preserved some of these artifacts, which were later given to the Museum.

Early Aircraft Collecting

Early in 1919 Lt. Col. Arthur Doughty (later Sir Arthur) went to Europe to collect war trophies. He brought back a large number of artillery pieces and similar weapons and a surprisingly good number of aircraft. These included the R.A.F. B.E.2c in which 2nd Lt. F. Sowery shot down Zeppelin L.32, and the fuselage of the Sopwith 7F.1 Snipe in which Col. W.G. Barker fought his epic VC-winning battle with about 60 enemy aircraft in October 1918. These were the only British aircraft acquired but Doughty managed to obtain at least 36 German aircraft, some of which were already with the Canadian Air Force at Shoreham, England, and others still on the Continent. This enemy aircraft collection comprised the following specimens: one Aviatik; one A.E.G. G.IV; four Albatros D.V. and D.Va; 21 Fokker D.VII; one Fokker E.V. (D.VIII); one Halberstadt C.V.; one Junkers J.I; four L.F.G. Roland D.VIb; one Phalz D.XII; and one Rumpler (likely a C.VII Rubild).

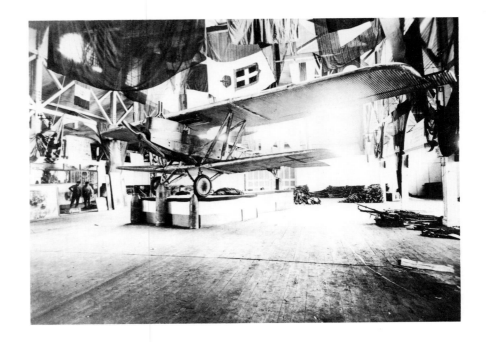

All these machines were shipped to Leaside Aerodrome, Toronto, in June 1919. Some were assembled, and the Junkers J.I and a Fokker D.VII were displayed at the Canadian National Exhibition in August of that year. Still others were displayed at Leaside, including Barker's Snipe fuselage and Sowery's B.E.2c. A number of Fokker D.VIIs were assembled and flown at Leaside, as was the Fokker E.V. Three or four D.VIIs were also used by a team headed by Col. Barker to give flying exhibitions at the Canadian National Exhibition and Barker flew one in the New York – Toronto Air Race.

After the CNE, the aircraft were shipped to Camp Borden, Ontario, and in 1920 several were sent as war trophies to a number of Canadian universities and cities as noted in Appendix 1. Lt. Col. R. Leckie, Director of Flying Operations, Canadian Air Board, ordered the German machines destroyed as he did not trust them and was concerned that pilots might be tempted to fly them. This distrust had been brought on by an accident at Shoreham caused by a structural failure in a Fokker D.VII, which resulted in the death of Major A.D. Carter. In fact, the D.VII was a good aircraft and continued to serve in several air forces during the postwar years. The Junkers J.I and A.E.G. G.IV escaped destruction, probably because they were unserviceable. In 1962 the Museum discovered the records in the Public Archives listing the shipping of some of these aircraft from Camp Borden and wrote the recipients to find out if any had survived, but none had. One engine existed and this was later obtained by the Museum for installation in their Fokker D.VII. As for the loss of this fine aircraft collection, it may be noted that the Canadian War Trophies Board was run by military men with no experience with or interest in aircraft, and while an annual check was made on the location and condition of all the rugged guns, there is no record of any check on the fragile aircraft.

Another collection had it origins in an extensive World War I training program which had been operated in Canada by the British RFC/RAF. Their School of Military Aeronautics at the University of Toronto had several types of service aircraft which were used for ground instruction to the cadets. At the end of the war three of these aircraft were turned over to the Royal Ontario Museum, operated then as part of the University. They were a Curtiss JN-4 (Can.), a D.H.4 and an R.A.F. F.E.2b; one of them, probably the D.H.4, was briefly displayed in the ROM. The three machines were sold in the early 1920s and the proceeds of the sale used to improve the University's telephone system. The name of the buyer was not recorded but it might have been a Mr. Holmes who had earlier inquired about their purchase. It is believed that the Curtiss JN-4 (Can.) and D.H.4 later came into the possession of the Central Technical School in Toronto and were used for ground instruction until the late 1930s. The F.E.2b may have been scrapped; alternatively it could have been sold in the USA, as one operated civilly there in 1922.

It seems probable that the Nieuport 12 now in the Museum collection arrived in Canada in the late 1920s or the early 1930s. The origin of the specimen is unknown. The first reference to it is in a letter of November 3, 1936, asking NRC to store it for the Canadian War Trophies Board and noting that "it is reported to have been a donation from the French Government." Later it was reported as being without fabric and without identification numbers. There is no record of it being shipped in 1919 along with the other war trophies.

Assuming that the Nieuport 12 was a French donation, it is interesting to speculate how this might have come about. Why would France suddenly decide to give Canada an obsolete aircraft? And, if so, why a Nieuport 12? The type, although used by several of the Allies, never distinguished itself and was disliked by its crews. The only plausible explanation seems to be one along the following lines. Suppose a Canadian with a non-aeronautical background became interested in the excellent performance of Canadian airmen in World War I and thought that Canada should have an artifact to commemorate its leading fighter pilot, Billy Bishop, and his VC-winning exploit. Obviously what was needed was a Nieuport 17, but knowing only that Bishop was flying "a Nieuport," which was how his aircraft was most often identified, this individual might have instigated a request for "a

Four Fokker D. VII fuselages being scrapped at Camp Borden circa 1921-22.

(KMM/G.R. HUTT)

John H. Parkin started the whole idea of a Canadian aviation museum, establishing a short-lived museum at NRC in the 1930s, and, with the help of others, a National Aviation Museum resulted in 1960.
(NRC)

Nieuport" from France, and France, trying to be helpful, sent a Nieuport 12.

Who might have had the interest and connections to have such a request made? Certainly one man who would fill all the requirements was Lt. Col. George A. Drew. His book *Canada's Fighting Airmen* was researched in the late 1920s and published in 1930. In writing of Bishop he refers to his machine as a Nieuport, never as a Nieuport 17, and it appears that he did not even realize that Nieuport made several different types of aircraft. Furthermore, Drew's political, military and social contacts were ideal to organize such a request to France. Be this as it may, all that is really known is that the Nieuport 12 was in Canada some time before 1936 and reported as a gift from France.

J. H. Parkin and the Aeronautical Museum

In the summer of 1929 an event occurred that had a profound effect on the establishment of Canada's aviation museum – John Hamilton Parkin was appointed to the National Research Council as Assistant Director, Division of Physics, and in charge of aeronautical research. It was Parkin and his extra-curricular efforts at NRC from then until 1960 that resulted in the formation of the Aeronautical Museum at NRC in the late 1930s and, then, the National Aviation Museum in 1960. While it was Parkin who spearheaded the effort to form an aviation museum, he had considerable assistance from other interested people in these early years.

Parkin was born in Toronto in 1891 and attended the University of Toronto, where he obtained a diploma in Mechanical and Electrical Engineering in 1911, a degree of Bachelor of Applied Science with honours in 1912 and a Mechanical Engineering degree in 1919. He worked at the university from 1912 onwards as demonstrator, lecturer, assistant professor and finally associate professor before joining the NRC. During this time he gave the first university courses in aeronautics in Canada and was in charge of the university's wind tunnel.

Parkin had been a member of NRC's Associate Committee on Aeronautical Research (known as the Associate Air Research Committee until 1931) since its inception in 1920 under its first two chairmen, Dr. A.S. Eve of McGill University and Wing Commander E.W. Stedman, Chief Aeronautical Engineer, DND. The purpose of the committee was to assist and promote aeronautical research in Canada and the name Associate Committee was chosen to indicate that it could include members, even its chairman, from outside NRC. Possibly the best known result of the Committee's efforts was the successful testing of W.R. Turnbull's electrically operated variable-pitch propeller in 1927, and the financing of the construction of the propeller by

Canadian Vickers. Another example was its assistance to Canadian Airways in service trials and the development of the Worth oil dilution system in the late 1930s.

It seems likely that Parkin got the Committee interested in promoting an aviation museum and collecting artifacts for it before he joined NRC, but certainly the Committee became more active in this field after his appointment.

Collecting Artifacts Between the Wars

On January 25, 1930, the first shipment of artifacts was sent to NRC for what was to become the Aeronautical Museum. It came from the University of Toronto and comprised eight aircraft engines of the World War I period together with some miscellaneous material.

Over the next few years donations of artifacts continued to come in, with the main donors being the University of Toronto and the RCAF, undoubtedly due to the influence of Parkin, G/C Stedman and, later, Maj.-Gen. A.G.L. McNaughton. Apparently the first exhibits started to appear in the halls of NRC in 1932. The exhibits comprised engines and were prepared by M.S. Kuhring and his staff at NRC's Engine Laboratory.

In December 1932 a Sopwith 2F.1 Camel was offered to NRC by the RCAF. This being an appreciably large artifact, Parkin sent Dr. H.M. Tory, NRC President, a memorandum on December 21, 1932. As this is the only document to outline Parkin's policy on collecting artifacts, it is presented in full:

Group Captain Ernest W. Stedman, Chief Aeronautical Engineer, RCAF (later Air Vice Marshal). G/C Stedman did a great deal to aid John Parkin and others in their efforts to establish an aviation museum in the 1930s, in particular by sending interesting artifacts to them.

(DND PL 117424)

"Re letter of December 21 from Deputy Minister of National Defence regarding a Sopwith 'Camel' aeroplane

"It is, of course, quite impossible for the National Research Council to provide the huge space necessary for a Museum to contain obsolete aircraft. The case of the 'Camel,' however, is of a special nature. It is the machine intimately associated with Canadian pilots during the war, and furthermore, it is understood that the machine referred to is practically the last of the type in existence. If it is not accepted by the Council, it will be destroyed. It is therefore recommended that, under these circumstances, the National Research Council should accept the offer of the department. When the machine is received, it is proposed to set it up as the central exhibit in the museum arranged over the Engine Laboratory.

"It is considered that the function of the National Research Council in connection with material such as the foregoing for museum purposes, should be that of accepting for storage such

The Sopwith 2F.1 Camel in its original RAF markings on display at the Aeronautical Museum in the National Research Council on Sussex Drive, circa 1937.
(J.S. BEILBY/NAM)

Dr. John J. Green obtained his PhD in aeronautics at London University in 1930 and immediately joined NRC in Ottawa. He became a most respected Canadian aeronautical scientist and in WWII he served with the RCAF as chief research engineer at Rockcliffe as well as being a test pilot. He made a significant contribution to the Museum.
(DR. J.J. GREEN)

pieces which would be otherwise destroyed. No other organization, apparently, has an interest in the preservation of such technical exhibits, and the Council would therefore be fulfilling a rather useful purpose in storing such pieces until they can be placed on exhibition in suitable quarters. Possibly in the future a National Science Museum may be established at Ottawa, and, if this is done, it would be the repository for such exhibits."

What Parkin's authority was for stating that the Camel would be destroyed is unknown but he was probably correct, for the aircraft had become the object of pranks by the flight cadets at Camp Borden. It was the only aircraft accepted by NRC for museum purposes and is the Camel now in the Museum collection.

It is interesting to reflect that while the historical significance of the Camel was recognized by NRC, other machines of special interest to Canada, such as the Canadian Vickers Vedette and Armstrong Whitworth Siskin, were aging and would soon be scrapped. What a pity that an unheated shed could not have been found and an example of each stored disassembled for the future.

In 1935 Maj.-Gen. A.G.L. McNaughton became President of NRC and a member of the Aeronautical Research Committee. He gave additional support to the Committee's efforts to preserve historical artifacts and in 1936 more space was allocated for their display at the NRC headquarters on Sussex Drive.

Also in 1936, the Committee made a successful effort to obtain some of the early aviation artifacts stored by the Bell family at Baddeck. This was instigated by J.A. Wilson, Controller of Civil Aviation, who had contributed much to the development of Canadian civil aviation. (Although not a member of the Aeronautical Research Committee, he was a strong supporter of its work in preserving Canada's aviation history.) Wilson wrote Maj.-Gen. McNaughton on February 26, 1936, to bring to his attention that some significant artifacts set aside by Dr. Bell were still held by the Bell family at Baddeck. He expressed concern over their present housing and suggested that if NRC would be prepared to house them better, it might be possible to bring some of them to Ottawa. The letter was followed on June 18 by a meeting in McNaughton's office attended by J.A. Wilson, F.W. (Casey) Baldwin, an original member of the A.E.A., and George M. Ross, Secretary-Treasurer of the Canadian Flying Clubs Association.

After seeing the NRC facilities, Baldwin agreed that some of the early artifacts at Baddeck should be turned over to NRC for preservation and display, and Wilson arranged for Civil Aviation Inspector Stuart Graham to fly Dr. J.J. Green, the Committee secretary, and his wife to Baddeck to talk with members of the Bell family. The party left Longueuil, Quebec, on August 10, 1936, in Waco YKS-6 seaplane CF-CCQ and flew via Saint John and Charlottetown to Baddeck, where they were guests of Baldwin and his wife.

The artifacts saved by Dr. Bell were housed in the original Kite House on the Bell estate. Mrs. David Fairchild and Mrs. Gilbert Grosvenor, Dr. Bell's daughters, agreed to donate the Curtiss engine of the *Silver Dart*; a cell of one of Dr. Bell's original kites; a *Silver Dart* propeller; and a family of his experimental model propellers, all of which were shipped to Ottawa that fall. Dr. Green has stated that the Bell family were most generous and that he felt they would have liked to have given more but he was reluctant to ask for more owing to the restricted space available at NRC. J.A.D. McCurdy also promised to donate his copies of the A.E.A. Bulletin recording the progress and accomplishments of the Association, eventually received in February 1938.*

The party left Baddeck on August 14 and stopped the next day at Rothesay, New Brunswick, to visit W.R. Turnbull who promised to donate the 1906 Duryea engine used in his early experiments, an early experimental propeller and a model of his variable-pitch propeller. In addition he said that he would try to arrange the return from England of his successful 1927 variable-pitch propeller. Unfortunately he did not succeed although the propeller was returned to Canada in September 1946 after Gen. McNaughton enlisted the aid of Professor Geoffrey Hill of Farnborough.

In 1937, without ceremony or publicity, an aluminum tablet was installed in the Museum at NRC. It was titled "The Aeronautical

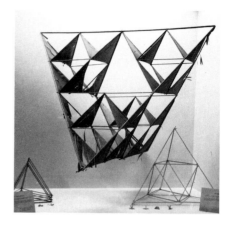

An original section of one of Dr. Alexander Graham Bell's tetrahedral kites.
(NAM 4005)

* The *A.E.A. Bulletin* and certain other publications donated are apparently still held by the NRC Library.

THE AERONAUTICAL MUSEUM
ORGANIZED UNDER THE AUSPICES OF THE
ASSOCIATE COMMITTEE ON AERONAUTICAL RESEARCH
1937
CHAIRMAN GROUP-CAPTAIN E.W. STEDMAN • DEPARTMENT OF NATIONAL DEFENCE
MEMBERS:
W.A. ACTON • DEPARTMENT OF TRANSPORT • • F.P. H. BEADLE • NATIONAL STEEL CAR CORPORATION
DR. R W. BOYLE • NATIONAL RESEARCH COUNCIL • • WING COMMANDER A.T.N. COWLEY • DEPARTMENT
OF TRANSPORT • • SQUADRON-LEADER A. FERRIER • DEPARTMENT OF TRANSPORT • PROF. T R LOUDON
UNIVERSITY OF TORONTO • • WING-COMMANDER C.M. McEWEN • DEPARTMENT OF NATIONAL DEFENCE
G. McINTYRE • IMPERIAL OIL LIMITED • • MAJOR-GENERAL A.G. L McNAUGHTON • NATIONAL
RESEARCH COUNCIL • • R.J. MOFFETT • CANADIAN VICKERS LIMITED • • J.H. PARKIN • NATIONAL
RESEARCH COUNCIL • • J. PATTERSON • DEPARTMENT OF TRANSPORT • • T.W. SIERS • CANADIAN
AIRWAYS LIMITED • • MAJOR H.E. TABER • DEPARTMENT OF NATIONAL DEFENCE • • M WINDSOR
ARMSTRONG SIDDELEY MOTORS LIMITED • • J. YOUNG • CANADIAN PRATT AND WHITNEY AIRCRAFT LIMITED

SECRETARY: DR. J.J. GREEN • NATIONAL RESEARCH COUNCIL
COOPERATING ORGANIZATION: • THE CANADIAN FLYING CLUBS ASSOCIATION
DONORS:
MRS. GILBERT GROSVENOR • MRS. DAVID FAIRCHILD • MR. F W BALDWIN
MR. J.A.D. McCURDY • • • MR. W.R. TURNBULL

Museum Organized Under the Auspices of the Associate Committee on Aeronautical Research" and it listed the Committee members and principal donors of historical artifacts. This is the first and possibly the only official use of the name "Aeronautical Museum."

In 1939 one of the two propellers hand-made in 1921 at Fort Simpson, Northwest Territories, for Imperial Oil's two Junkers JL-6s was presented to the Museum by Father Gathy, O.M.I. The second propeller was tracked down by Frank H. Ellis, Canada's first aviation historian, and was given to the Museum in 1945 by its owner, Mrs. G.W. Gorman, widow of one of the pilots. Frank Ellis was also responsible for tracking down W.W. Gibson, who had made Canada's first aircraft engine, which he donated to the Museum in 1944.

Meanwhile the outbreak of World War II necessitated the closing of the Museum at NRC and all artifacts were put into storage. In 1944 a 100 hp air-cooled Vee engine stated to have been used by Count Zeppelin was offered to the Museum. It was turned down as "not being Canadian." However, Count Zeppelin never used air-cooled engines so what was turned down will never be known. Disaster nearly overtook the whole collection in 1945 when it was suggested that "the old engines and stuff" in storage should be disposed of. Fortunately Parkin heard of the idea and stopped it. This illustrates the danger of leaving artifacts in the custody of an organization whose principal purpose is

20

not their preservation. Sooner or later they will be in the way and may be destroyed.

Efforts to Resurrect the Museum

After the war, the RCAF decided to save specimens of its wartime aircraft. It is not known just who made the decision or when it was made, but it may have been influenced by General H.H. "Hap" Arnold's instruction to the USAF to "save one of each." It is also possible that the Canadian War Museum's 1945 decision to display full-size aircraft may have played a major part. Unfortunately no effort was made to save any of the RCAF's between-the-wars types which are known to have been in existence at that time.

The Aeronautical Research Committee held its last meeting in 1948 and little happened in the way of resurrecting the former Aeronautical Museum or creating a replacement until 1954. At the annual meeting of the Air Industries and Transport Association (AITA) in 1950 the matter of a museum had been raised by George Drew. It was again discussed at the 1953 annual meeting and the idea was commended by Prime Minister Louis St. Laurent, who was present. The matter was apparently again raised at a subsequent AITA meeting by the AITA Secretary A.G. Sims, and after a brief discussion was referred to the Information Committee to be dealt with at their next meeting. The AITA Information Committee met on June 16, 1954, at Canadair Ltd. with A.G. Sims as Chairman. After considerable discussion it was decided to contact J.H. Parkin and Dr. J.J. Green and to approach interested organizations to determine their support.

As a result, a meeting was called at Ottawa on September 20, 1954. Present were: Gen. A.G.L. McNaughton, Chairman, Permanent Joint Board on Defence; Air Marshal C.R. Slemon, Chief of Air Staff; Dr. E.W.R. Steacie, President, NRC; Dr. O.M. Solandt, Chairman, Defence Research Board; R.N. Redmayne, Manager, AITA; J.A. Wilson, Director of Air Service, DOT (retired); J.R. Baldwin, Deputy Minister, DOT; Parkin who was then Director of Mechanical Engineering at NRC; Sims, Secretary, AITA; and Dr. Green, Chief Division "B", Defence Research Board. A/M Slemon acted as Chairman. It was decided that a national aviation museum was required, that it should be located in Ottawa and that an NRC Associate Committee should be formed to further its establishment. It is doubtful that a more qualified and influential group could have been assembled to deal with this matter.

The formation of an NRC Associate Committee on a National Aviation Museum began almost immediately but its first meeting did not take place until December 5, 1955. At that time the members were Gen. McNaughton, as Chairman; Dr. Green, Defence Research Board;

W.W. Gibson's engine on display at the Aeronautical Museum in 1949. This photo indicates that items were displayed, at least briefly, in the early post-WWII period. It is also, possibly, the sole photograph showing Aeronautical Museum exhibits. (NRC)

Arthur G. (Tim) Sims, former air engineer, bush pilot, RAF Ferry Command pilot and Canadair's Director of Military Aircraft Sales. He raised the whole matter of forming an aviation museum in 1953.
(BASIL ZAROV/VIA MRS. A.G. SIMS)

Malcolm S. (Mac) Kuhring, Head of the Engine Laboratory, NRC, was small of stature but overflowing with energy; it was his role to bring into fruition the ideas of the two museum committees. The Museum owes much to his efforts.
(NRC)

E.R. Hopkins, RCAF Association; J.R.K. Main, Department of Transport; W.P. Paris, Royal Canadian Flying Clubs Association; J.H. Parkin; Air Vice Marshal J.L. Plant, Department of Defence; G.M. Ross, Air Cadet League of Canada; A.G. Sims, Canadair Ltd.; J.M. Manson, NRC, as Secretary; and E.A. Côté, Deputy Minister, Department of Northern Affairs and National Resources, who was not present at the first meeting. Dr. J.J. Green succeeded Gen. McNaughton as Chairman in April 1965 and H.C. Luttman, Canadian Aeronautical Institute, became a member in 1957.

At the first meeting of the new committee, the Department of Transport offered space in the new Uplands Airport Terminal Building. An offer by A.V. Roe Canada Ltd. to donate its C-102 Jetliner was discussed and turned down due to lack of space. However, a Chinook engine was requested from the company.

It was May 9, 1957, before another meeting of the Museum Committee took place. Parkin was discouraged about the situation and presented a brief on an engineering museum. Luttman announced that the Canadian Aeronautical Institute was planning to make a full-scale reproduction of the A.E.A. *Silver Dart* for the 50th anniversary of its first Canadian flight. The Committee was pleased with this and expressed interest in receiving the reproduction at a later date. The Committee learned that Shell Oil Ltd. and Imperial Oil Ltd. would consider an appeal for funds for the Museum.

The third meeting took place on November 25, 1957, and the Department of Transport stated that the space in the new Uplands Terminal Building would no longer be available. There was some discussion of placing the available artifacts on display at various unnamed airport terminal buildings. Luttman reported that as the RCAF had completed an A.E.A. *Silver Dart* reproduction and intended to fly it at Baddeck on February 23, 1959, the CAI had decided to discontinue its project.

At a fourth meeting on October 25, 1958, it was reported that space in the new Uplands Terminal Building would now be available, and a sub-committee with M.S. Kuhring as Chairman was set up to examine the space and report back. Gen. McNaughton and E.A. Côté had prepared a draft memorandum on the Museum to be presented to Cabinet by the NRC President upon approval. Not mentioned at the meeting was the fact that a brief on the Museum had already been presented to the Prime Minister on July 8, 1958. An opening date for the Museum was discussed.

In the meantime, the Museum's collection continued to grow and develop. In December 1957 a very well-sectioned, electrically-powered Napier Sabre engine arrived from its makers, and in 1958 the Avro Canada Chinook engine arrived, the first Canadian jet engine. In 1958 the RCAF arranged to borrow and restore the Nieuport 12, R.A.F.

B.E.2c and Sopwith 2F.1 Camel at 6 Repair Depot, Trenton. In 1959 the B.E.2 and the fuselage of Barker's Snipe were displayed in the Canadian War Museum and the Camel and Nieuport 12 were displayed at Trenton.

On February 12, 1959, the Sub-Committee on Space reported to the Museum Committee that 1211 square metres would be available on the second floor of the east wing of the Uplands Terminal Building, together with space in the display area downstairs at the front of the building for the *Silver Dart* reproduction, which had been promised by the RCAF. It was agreed that the Museum Committee should continue in an advisory capacity after the Museum opening which was now scheduled for July 1, 1959, although there was some doubt that the necessary work

John Dost, Assistant Curator, and an unidentified man examine the Avro Canada Chinook, Canada's first jet engine.
(NAM)

23

could be done by then. Funds were received from Fleet Aircraft Ltd., Bristol Aeroplane Co., and Rolls-Royce of Canada and a promise from Shell Oil Ltd. A Sub-Committee on Finance was established composed of J.H. Parkin and E.A. Côté. M.S. Kuhring was appointed Curator *ad interim*.

The idea of opening the new Museum in 1959 was, from the outset, an optimistic one. The events of August 5 scuttled the idea completely. On that day, Capt. George L. Schulstadt, USAF, was demonstrating a Lockheed F-104A at Uplands Airport. During a low pass, the fighter, apparently inadvertently, became supersonic, causing a sonic boom that did severe damage to the almost-complete Terminal Building, scheduled to open in September. A hundred or more glass panels, some as large as 5.5 by 2 metres, were broken and the ceilings on all three floors were damaged. After repairs the new terminal came into use on June 15, 1960, and was officially opened on June 30.

At the next meeting of the Museum Committee held on November 12, 1959, M.S. Kuhring presented a model of the proposed layout. It was reported that Pacific Western Airlines had the last remaining Junkers W.34 and that it could probably be donated but it was felt that complete aircraft could not be accommodated. Kuhring was asked to try to have PWA hold the Junkers until the Museum was more fully developed. This, of course, was the same Junkers W.34 that was presented in 1962 by Mrs. J.A. Richardson. The writer was appointed to the position of curator reporting for duty at the beginning of July 1960. Three other staff members were also appointed at that time.

At first all we could do was to follow the energetic, fast-moving "Mac" Kuhring and try to understand what was planned and what was being done. The layout of the Museum was easy enough to follow as a good plan existed and construction was well advanced. The displays were something else again though, as plans for most of them existed only in Mac's fertile mind. After six or seven weeks I was given responsibility for the design of four model display cases which had not yet been planned. By working overtime we managed to have all major items complete a full day before the opening and the staff were able to spend a leisurely day checking over details instead of the mad scramble that often takes place on such occasions.

Inauguration of the New Aviation Museum

The opening of the new National Aviation Museum by Prime Minister John Diefenbaker was set for October 25, 1960, at 2.30 p.m. Unfortunately, pressing duties prevented the Prime Minister from attending and Fisheries Minister J. Angus McLean performed the official opening.

Following this, Gen. McNaughton gave the opening address to the guests, among whom were J.A.D. McCurdy, Mrs. F.W. Baldwin, Donald Turnbull, Air Marshal R. Leckie, Air Vice Marshal A.E. Godfrey and Mrs. Gilbert Grosvenor, together with many others who had contributed to the Museum and to Canadian aviation history. After the ceremony in front of the *Silver Dart*, the guests proceeded upstairs to see the displays. It was a very satisfying day for all those who worked so hard to bring the Museum into existence.

The following day the Museum staff was on its own for the first time and Mac Kuhring and the others from NRC went back to their normal duties. Before passing on to new events in the Museum's history, a word must be said about Mac Kuhring. Unless one had the privilege of working beside him, one cannot really appreciate how much hard work he put into the creation of the Museum, certainly far beyond the call of duty. Indeed it is hard to know how it all could have been done without his efforts. In appreciation of his work and his fine contribution the Museum Committee presented him with a silver cigarette box at their meeting on November 9, 1964.

Improving Museum Exhibits

The first task of the NAM staff following the opening was to organize their workshop area and set up their tools, for all prior efforts had understandably been directed to the exhibits. It soon became apparent that storage space for shop material had been overlooked in the layout and new storage space had to be created by gaining access to unused areas behind or beneath exhibits. This practice continued until the Museum vacated the Terminal Building. A visiting official once remarked that he had never seen space so efficiently used.

It soon became apparent that there was much to be learned about the design and construction of exhibits. Many of the exhibits that could be operated by the public had to be reworked to withstand the rough treatment they received. This work consumed most of the staff's time for some months after the opening, and the experience was used to advantage in designing new exhibits.

As soon as the workshop was set up and the exhibits put in order, a program was begun to upgrade the original exhibits. The display cases in each area of the Museum had been assigned a separate colour and the artifacts had been set up in them with little or no attention to design or colour co-ordination. The one exception was the four model display cases mentioned earlier. In that case a local artist had been engaged to prepare coloured sketches of each display, which were then translated into three-dimensional exhibits by the staff. The result was satisfactory and this procedure was followed during the entire stay at

Left to right, the nose of a Canadian Vickers Vedette, aircraft skis, cut-away rotary and radial engines, model display cases, and a D.H.60 aircraft float.

(NAM 3828)

the Terminal Building. The result was a gradual brightening up of the original exhibits along with attractive new ones.

New exhibits were created from time to time. Owing to floor loading restrictions imposed by the second-floor location, heavier exhibits, such as the operating cut-away specimen of the Avro Canada Orenda jet engine had to be designed with large bases to distribute the load.

From late 1962 to 1966 the Museum attempted to restore two or three aircraft engines in its collection each year for display on suitable stands in the Museum Annex, where more space had been acquired. In addition to expanding and improving the exhibits, there were several other important areas of museum operations to be developed. These were a historical aviation library, a photograph collection and an aircraft model collection.

Donations Help Create the Library and Photo Collection

It was essential that the Museum have a good library of books, reports, manuals and pamphlets for use as reference material in setting up exhibits, repairing and overhauling aircraft and engines, making models, and other Museum operations. At the beginning there was no library in Canada devoted to aviation history. The National Research Council had a good technical reference library which served Canadian engineers well but it had only a limited amount of historical reference material.

A program was started to set up a library. It has, in hindsight, been more successful than anyone might have predicted in 1960. Old, out-of-print aviation books are difficult and expensive to obtain and, of course, the better and more useful ones are quickly snapped up when they become available. Generous donations, hard work and some good

National Aviation Museum staff, May 1964, left to right:
J. Dost, Assistant Curator;
K.M. Molson, Curator; A. McRorie;
O. Clarke; A. Walker; G. Telford;
W. Merrikin; unidentified
temporary helper.
(NAM 2-20)

luck have resulted in the Museum having the best historical aviation library in Canada. Many individuals have donated books over the years, and the late N.E. Butler and the late A.G. (Tim) Sims both gave sizable collections.

The Institute of Aeronautical Sciences in New York was approached to see if they might have some duplicates in its fine historical aviation library. This proved to be the case and they generously agreed to make two donations amounting to about 600-700 volumes. Many were old and difficult or almost impossible to obtain.

Late in 1964 the Museum was fortunate to be able to purchase the library of the former Austro-Hungarian Aero-Technical Society of Vienna, which was for sale in Europe. It contained a large number of books and periodicals and was especially rich in pre-World War I material, which is very scarce. This purchase and the IAS donation were responsible for a large part of the Museum's present library. Helpful donations continue to be made and together with purchases, the library continues to grow.

An aviation photo collection was badly needed for several reasons. Canadian books and articles often featured foreign photos for illustrations because there was no source in Canada other than the Department of National Defence. A check of the Public Archives in 1961 revealed only two aviation photos. The Museum uses photographs for many purposes – as background material for exhibits, for historical reference, for reference in constructing models and, probably most important, in restoring aircraft specimens. Pictures that would not win any photographic prizes are frequently most useful for historical reference. Photographs proved to be especially valuable in the restoration of the Curtiss JN-4 (Can.) and the Curtiss HS-2L, yielding invaluable information not otherwise available. The collection has been built through loans and donations of photographs and it continues to grow with the help of thoughtful people.

Establishing the Model Collection

An aircraft model collection is an almost essential part of an aviation museum. Actual specimens are simply not available to portray all the progress that has been made in aviation, and even if they were available no museum would have the space to display them. Photographs are not always satisfactory as they are two-dimensional and the aircraft cannot be viewed from different angles to discern all their features. Models also show colour and when well displayed can be viewed from different angles so that all details become visible.

Great care is required to produce a first-class museum model. Good aircraft modelmakers are few, and most often are craftsmen, not historians. It is necessary, therefore, to provide drawings, photographs and colour information to define the aircraft completely in all required details. Finding all the necessary information and checking it for accuracy can be difficult and time-consuming. Published drawings are sometimes inaccurate and photographs of some types can be difficult to obtain. In some cases it is necessary to prepare the drawings. Inevitably the making of museum models is time-consuming and expensive, and like most other things, they have increased greatly in price in recent years.

Museum models should be made to a single scale so that they all appear in the correct proportion to one another. There is no universally accepted scale but they should be large enough to show good detail, yet not so large that the models of bigger aircraft become difficult to display. The standard for the Museum was established as 1/24 scale, or 1/2 inch equals one foot. This scale was already in use by several other museums and was also adopted by the Canadian War Museum. Scales at other museums vary from the 1/10 scale of the Musée de l'Air in

*Aircraft Models
from the
National Aviation
Museum Collection*

A high quality model collection is indispensable for an aviation museum. Models permit the visitor to see close up, in three dimensions, aircraft not on display or not in the museum collection. From different angles, all details are visible, especially the aircraft colours and markings.

The four-engined Sikorsky Grand was first flown in May 1913 at St. Petersburg (now Leningrad), Russia. It was the ancestor of all later four-engined transport and bomber aircraft. (KMM)

The Armstrong Whitworth Siskin IIIA fighter was introduced into the RCAF in 1926 and was its only fighter until the late 1920s. It is best remembered for outstanding aerobatic displays in the late 1920s and early 1930s. (KMM)

The Ford trimotor family was the best known and longest lived of American transport aircraft of the 1920s. Only a few examples of the "Tin Goose" saw Canadian service and the Ford 4-AT modelled here, G-CATX of British Columbia Airways, became the second of the Canadian Fords in August 1928. (NAM)

In 1931 Supermarine S-6B piloted by F/L J.N. Boothman, RAF, won permanent possession of the Schneider Trophy for Britain in 1931 at a speed of 340.8 mph (548.4 km/h). (NAM)

The National Aviation Museum has a fine collection of models. It features interesting Canadian machines, significant foreign aircraft, WWI aircraft, some high-speed and long-distance record-breaking machines, and transport aircraft. Illustrated below is a selection from the Museum's collection. All models are crafted in the standard 1/24 scale.

The Douglas MO-2BS was a civil seaplane version of the US Air Service O-2B observation machine made to J.D. McKee's order. In 1926 McKee and S/L A.E. Godfrey, RCAF, made the first seaplane flight across the North American continent from Montreal to Vancouver. Shortly thereafter, McKee donated the Trans Canada Trophy (McKee Trophy) as an annual award to a person making an outstanding contribution to Canadian aviation.   (KMM)

The famous Bellanca WB-2 Columbia, later Maple Leaf, was the first aircraft to complete two Atlantic flights. The first one was in 1927, immediately following Lindbergh's flight, the second in 1930 with Canadian pilot Errol Boyd, who was the first Canadian to make the crossing. This aircraft was the first of a distinguished Bellanca family of long-distance record breakers.   (KMM)

The Fairchild Super 71, designed and built at Longueuil, Québec, was the first Canadian machine designed for freighting in the bush. It was a good freighter but only one was built because its all-metal fuselage made it more expensive than its contemporaries, and some pilots objected to the rear cockpit position.   (KMM)

The Avro Canada C-102 Jetliner was the first Canadian-designed jet aircraft and became the second jet transport to fly on August 10, 1949, at Malton, Ontario.   (NAM)

France, which produces very large models, to the 1/72 scale used by some smaller museums.

The models present on the opening of the NAM were all made by amateurs, with two exceptions, and had to be replaced as soon as possible. Aircraft modelled for the Museum in subsequent years were almost all machines made before 1939 and fell into the following categories: significant or interesting Canadian machines, significant foreign aircraft, representative aircraft of World War I, high-speed and long-distance record-breaking aircraft, and transport aircraft. Few models could be obtained in any one year because of the cost but some of each class were obtained annually until 1967.

A model program of a different nature was started in 1964 to demonstrate various aspects of aircraft construction and operation. The first was a series of models showing different types of wing construction; a second series of working models showed different types of undercarriage retracting mechanisms. Others were planned but the program stopped on the absorption of the NAM by the National Museum of Science and Technology in 1967. This was a pity, for the completed exhibits brought much favourable comment from visitors.

Aviation Curators Form an Association

Once the Museum had settled into its new quarters I felt that, since I had no aviation museum experience other than as a visitor, it would be a good idea to meet with people who did. A brief visit to the National Air Museum (later the National Air and Space Museum) in Washington in the spring of 1961 led to close contact with Louis S. Casey, Curator of Aircraft, who visited Ottawa in 1962. These reciprocal visits continued. As the problems and interests of aviation museums are quite different from those of other museums, Casey suggested a group of aviation museum curators be formed. A meeting of twelve directors and curators at Washington in April 1966 agreed that an association should be formed, and meetings have been held each year since at museums in places as far apart as Pensacola, Florida; Hartford, Connecticut; San Diego, California; and Ottawa. Two meetings have been held in Europe. Papers are presented and there are tours of the host's workshops and exhibits. While all this has been useful, the most valuable single result is that the curators now know one another well, and this has proved a great, if intangible, asset.

In the early years of the Museum's existence there was concern that there was no visual historical record in colour of early Canadian aircraft, historic Canadian aviation events and sites. To rectify this, a series of paintings to provide accurate and pleasant reconstructions was planned. The problem was to find a suitable artist. The subject

matter requires a firm command of perspective, and the artist had to have some aviation knowledge, preferably first-hand, in order to present correctly a machine's various positions in flight. In addition the artist had to have the skill to portray the locales from which the machines operated.

The Museum was most fortunate to find such an artist in Robert W. Bradford. He agreed to work on a series for the Museum and his resulting fine work brought him recognition as a leading aviation artist. It was hoped that four paintings a year could be completed but this target was never reached before the project ended in 1967. A variety of subjects has been portrayed, including the first flight of the A.E.A. *Silver Dart* at Baddeck; two fine scenes of the first trans-Canada flight in 1920; John Webster flying his Curtiss-Reid Rambler III in the King's Cup Race; the Canadian Vickers Vedette over its birthplace, and many others.

When the National Aviation Museum was first set up, it came under the Department of Northern Affairs and Natural Resources. That this had been intended for some time was indicated by the presence of the Department's Deputy Minister, E.A. Côté, on the Associate Committee for a National Aviation Museum from its inception in 1955. The rationale for not including the Museum under the Secretary of State Department like the other national museums at that time is not known, but it was moved there in 1961 and reported through the Director of the Museum of Human History, Dr. Loris Russell, and his successor, the late Dr. R.G. Glover.

On display at the Museum Annex, left to right, are the Curtiss JN-4 (Can.), D.H. 60 Moth and Nieuport 17; behind the latter is the A.E.A. Silver Dart. *Between the JN-4 and D.H. 60 is a Curtiss OX-5 engine.*

(NAM 4564)

W/C Ralph V. Manning, DFC, joined the RCAF soon after the outbreak of WW II, served overseas with the RAF in Nos. 42 and 47 (Torpedo Bomber) Squadrons and for his distinguished service was awarded the DFC in 1944. He did much to augment the RCAF's historic aircraft collection and played a prominent role in obtaining the Rockcliffe site for the three historic aircraft collections then in existence.

(DND PL 1049989)

In the years following its opening, the Museum's display space and other facilities at the Uplands Terminal expanded considerably. At the beginning there was space to display only a single full-scale aircraft. In 1963 the Museum obtained the use of the whole of the downstairs display area at the Terminal Building, which became known as the Museum Annex. It was then possible to display four aircraft together with several engines. One or more aircraft were changed each year and specimens that appeared during the early period included the A.E.A. *Silver Dart*, Curtiss JN-4 (Can.), D.H.60 Moth, Fairchild FC-2W-2, Junkers W.34, Nieuport 17 and Supermarine Spitfire IX. After the NAM was absorbed into the National Museum of Science and Technology in 1967, some aircraft continued to be displayed there until the 1970s, including the D.H.60 Moth, Lockheed 10A Electra and Fairchild 82A.

The Collections Move to Rockcliffe

In 1959 Wing Commander Ralph V. Manning, DFC, was appointed RCAF Air Historian to succeed the respected W/C F.H. Hitchins, who retired in 1960 after 19 years of work in Canadian military aviation history. W/C Manning, who had had a distinguished career in World War II flying Bristol Beauforts, brought to the assignment an interest not only in RCAF history but in historical aircraft and their preservation. He had been anxious for some time to obtain a place to display some of the RCAF historical aircraft and was also conscious of the growing number of aircraft held by the Canadian War Museum and the National Aviation Museum. At one time he considered using Victoria Island on the west side of Ottawa, which was to be vacated by the RCAF, but he gave up the idea, probably because it would not have been suitable. Manning's efforts to preserve and display historical aircraft were many but probably the most significant was his role in helping bring about the allocation of the Rockcliffe hangars in 1964 for housing Museum aircraft. In addition he made a determined effort to collect from private sources historical aviation photographs, something that had not been done before by the RCAF and that has proved most useful. In carrying out this work he had the willing support of numerous RCAF officers including A/C G.G. Diamond and W/C W. Pearce.

Until 1964 most of the RCAF's historical aircraft had been stored and displayed at Mountain View, Ontario, a satellite field of No.6 Repair Depot, Trenton. Some were held in western Canada. The considerable man-hours spent in erecting, removing and returning these machines for temporary display each year became an irritant to the commanding officer. S/L J.L. Murphy approached W/C P. de L. Markham with the problem. W/C Markham had taken an interest in the historical aircraft and was aware that the Rockcliffe base would be coming available. He submitted the idea of moving the aircraft collection to Rockcliffe to A/V/M J.B. Millward, who approved follow-up action. This brought in W/C Manning, who handled the matter at Air Force Headquarters. A quick decision was not obtained, but to forestall another display season at No.6 RD, W/C Markham obtained permission to move the historic aircraft to Rockcliffe in spring 1964. Some refurbishing work was carried out by No.6 RD on the aircraft before the move, and some fire prevention work was carried out on the Rockcliffe hangars also.

Upon the RCAF abandoning the use of the Rockcliffe airfield in 1964, an agreement was made between the Secretary of State Department, (which was responsible for the Canadian War Museum and the National Aviation Museum), and the Department of National Defence concerning the use of the Rockcliffe facilities to house all the aircraft held by the three organizations at one place, except, of course, for those specimens already on display at the Canadian War Museum and in the Terminal Building at Uplands. This brought into being a sizable collection of which Canadians could be very proud. The agreement called for joint management of the three collections at Rockcliffe on an experimental basis.

The assembly of these machines at Rockcliffe made the last Air Force Day display there on June 4, 1964, a "bang-up" affair. Air Force Day displays were put on at a number of air stations but, Ottawa being the capital and the site of RCAF Headquarters, the Rockcliffe displays were special. Rockcliffe's topography made them a treat for the favoured invited guests who sat on chairs on the lawn in front of the Officer's Mess, located on the top of the cliff on the south side of the airfield. The guests thus overlooked the field where the displays took place close to eye level against the backdrop of the Gatineau Hills in the distance. Behind the guests was a bar to slake the thirst brought on by the warming rays of the sun (always shining on these occasions). An RCAF band provided music and it is unlikely that any air display was ever watched under more pleasant conditions. It is not surprising that an invitation to the Air Force Day Display was said to be coveted in diplomatic circles.

After the air force display, the aircraft of the three Rockcliffe collections remained on public exhibit through the summer until the fall of 1964 when Rockcliffe closed for the winter. On the evening of May 21, 1965, an official opening of the exhibit was held and the term National Aeronautical Collection was used officially for the first time to embrace all the specimens of the three collections. The Minister of National Defence, Paul Hellyer, officiated at the ceremony.

The staffing arrangement from 1964 to 1967 was that the CWM would provide staff under J.F. Murphy to carry out the daily housekeeping duties, while the NAM restoration staff moved in to carry on their restoration programs as well as pitching in when required to assist the CWM staff. The arrangement worked well, as Squadron Leader Jack Murphy, RCAF (retired), was an ideal person to carry on the necessary liaison with the NAC's landlord, the RCAF.

Flying operations began at Rockcliffe in the summer of 1920 with the Air Board Avro 504K flown by C.M. McEwen.
(DND HC 33)

Rockcliffe – The Ideal Location

There is no doubt that the events of 1964 were the deciding factor in making Rockcliffe the permanent location for the Museum, as it has had a long and interesting aviation history.

Rockcliffe Aerodrome and Seaplane Station was established on a former rifle range by the Air Board in 1920 for their civil air operations. It was the only combined aerodrome and seaplane station at the time and in those early days was one of only two Air Board aerodromes, although the Canadian Air Force had another at Camp Borden. The first Air Board operations from Rockcliffe were in connection with an aerial survey of Ottawa, which was started in an Avro 504K flown by C.M. McEwen.

With the formation of the RCAF in 1924, Rockcliffe became its only combined aerodrome and seaplane station until Trenton Air Station opened in 1931. In 1925 all water operations were transferred to Shirley's Bay, west of Ottawa on Lac Deschênes, but this was not satisfactory and they were returned to Rockcliffe in 1929. In the meantime Rockcliffe had been improved by the addition of a new

hangar on the north side of the airfield and by a new slipway. In 1935 a photographic building, which still stands, was built on the south side of the airfield. In 1940 the hangars later occupied by the Museum were built.

Most of the flying at Rockcliffe was done by the Air Board and the RCAF, but there were some interesting civil operations too. In 1920 the first trans-Canada flight, carried out by the Air Board, stopped there on its way west. The noted American speed flyers F.M. Hawks and J.R. Wedell both flew from there in making new records. Col. Charles Lindbergh and his wife stopped there on July 30, 1931, on their flight north to the Orient, and J.R. Grierson landed there on August 30, 1934, in his D.H.83 Fox Moth seaplane on his flight from England via Iceland and Greenland. A tragedy also occurred there on March 12, 1930, when W/C W.G. Barker, VC, was killed at Rockcliffe in a flying accident.

Until 1938 all new Canadian aircraft were tested at Rockcliffe for airworthiness, and any new machines being demonstrated to the RCAF also appeared there. During World War II the RCAF's Test and Development Establishment was located at Rockcliffe, so numerous new aircraft were tested there, along with modifications to existing

The Museum has taken delivery of a number of aircraft by air since 1962, including seaplanes, fighters and machines as large as the Canadair CL-28 Argus. The first and oldest was the Junkers W.34 shown here. It arrived on September 17, 1962, after a ferry flight from Vancouver, B.C.

(NAM 3808)

types. After the war, the T & D Establishment moved to Uplands Airport. No. 168 Squadron operated from Rockcliffe from October 1943 to April 1946 carrying mail overseas to Canadian forces in Europe in Boeing B-17 Flying Fortresses and Consolidated B-24 Liberators. After the RCAF ceased flying operations at Rockcliffe, the airfield continued to be used by the Rockcliffe Flying Club and it was used by Air Transit for their trial STOL service to Montreal from July 1974 to April 1976, the only passenger service to operate from Rockcliffe. The terminal building and hangar erected for the service were taken over by the Museum, which has used both the airfield and seaplane facilities for taking delivery of aircraft specimens; the first of these was the Junkers W.34 seaplane in September 1962 and the most recent at the time of writing was the Lockheed Jetstar on June 9, 1986.

Consolidation of the Collections

To lead up to the consolidation of the three collections under the National Museum of Science and Technology in 1968, one should go back to 1880 and the founding of the Military Museum, which became the Canadian War Museum in 1942 as apparently some felt the earlier name implied that it was not interested in naval or air force matters. The Canadian War Museum, like its predecessor, was controlled by a board on which the three services were represented. Consequently, the CWM was thought of as the historical representative of the three services in the museum or artifact area.

In its early years the CWM occupied a very small facility and had a small staff. Its collections were confined to small battle trophies, clothing and personal equipment, field equipment, small arms and so on. In 1942 it began to collect larger items, but full-size aircraft were thought to require too much space so scale models were to be collected. A program was set up to have Canadian aircraft manufacturers supply models of their products. By 1945 the thinking had changed and the CWM now proposed to enlarge its facilities to include full-scale artifacts, including aircraft. The enlarged facilities were not obtained until 1967 but the new policy may have been what initiated or strengthened the RCAF's idea of setting aside some specimens of their World War II aircraft.

In 1951 the RCAF advised the Canadian War Museum that it was holding 23 aircraft of 13 types for the CWM's Board and asked for guidance on what the CWM would require. The response is not known, but both Avro Anson IIs, Bristol Bolingbrokes and Airspeed Oxford Vs and the single Messerschmitt Me 262 must have been discarded, as they and seven duplicate specimens have not survived. About 1960 the RCAF stated its intention of transferring their historical aircraft to the

CWM, a program which started with the transfer of the North American P-51D Mustang IV in December 1961. The program ceased when it appeared that the CWM did not have adequate facilities to look after the aircraft.

Also about this time, the RCAF Association was collecting funds for an RCAF memorial at Trenton which would include the display of some historic aircraft. It seems likely that the transfer of machines to the CWM was held in abeyance pending the finalization of the plans for the Trenton memorial. While considerable funds were raised, they were not sufficient to proceed with the Trenton memorial and have been used to create an RCAF place of tribute within the new Museum building.

In June 1965 a conversation was held between Dr. R.G. Glover, to whom the War Museum and the NAM reported, and C.P. Stacey, Director, Historical Section, Canadian Forces Headquarters, concerning the National Aeronautical Collection. Dr. Stacey was reported to have said that "the sooner the responsibility for the NAC can be concentrated in one authority, the better it will be. The present situation seems to me awkward and likely to lead to difficulty. The National Museum would seem to me to be the place where authority should ultimately be entirely concentrated." However, he went on to say the opinions he gave were not the settled policy of the Department of National Defence.

In the fall of 1966 Dr. David M. Baird was appointed Director of a new National Museum of Science and Technology, and the National Aviation Museum was assigned to him.

In December of the same year Dr. Baird raised the question of responsibility for the National Aeronautical Collection with the Under Secretary of State, G.G.E. Steele, who convened a meeting with the two museum directors concerned, Dr. Baird and Dr. Glover, along with representatives of the Canadian War Museum, National Aviation Museum, Department of National Defence and the RCAF Historical Section. The meeting was held because the experimental period of the National Aeronautical Collection was over and it was time to review its administration and development.

As a result of the meeting Mr. Steele put forward the following comments and proposals to the DND for consideration which, briefly stated, went as follows: The National Aeronautical Collection had been a great success, and, in view of the great Canadian involvement in aviation, it seemed desirable to ensure that the historic aircraft collection had the best possible display. Considering resources, it seemed desirable to have a single excellent museum rather than several smaller or less successful ones, and, it was, therefore proposed that the National Aeronautical Collection be placed under the National Museum of Science and Technology.

The Sopwith 2F.1 Camel has been on display at the Canadian War Museum since 1967. In the background is the fuselage of Col. Barker's famous Sopwith Snipe.
(NMC K83-179)

In some quarters of the DND there was opposition to the idea of their military aircraft being displayed with civil machines and technological exhibits. It was even suggested that Rockcliffe be made a purely military aircraft museum and that the NAM confine itself to civil aircraft and technological exhibits. However, in due course the original proposal was agreed to and the National Aeronautical Collection came under the direction of the National Museum of Science and Technology in 1968. This did not mean that the CWM would not display aircraft but simply that the main aircraft display would be at Rockcliffe and that the National Museum of Science and Technology would be responsible for all historic aircraft, their housing, upkeep, restoration and so on. The name National Aeronautical Collection was retained.

In the spring of 1967, the exhibits of the National Aviation Museum were moved from the Uplands Terminal Building to the National Museum of Science and Technology, now located in its new quarters at 1867 St.Laurent Boulevard. The new museum was officially opened on November 15, 1967. Also in 1967 it was agreed that some aircraft of the National Aeronautical Collection would be flown in a special June 6 display to help celebrate Canada's Centennial. The RCAF provided a team of pilots composed of F/Ls N.A. Burns (Aeronca C-2), J.F. Fitzgerald (Sopwith Triplane), W.R. Long (Nieuport 17), J. MacKay (Fleet 16B)

40

National Aviation Museum staff,
July 8, 1986, left to right:
Barry Mackeracher, Philippe Roger,
Kelly Cameron, C.F. (Chuck) Aylen,
William Merrikin, Joe Dorn,
John Bradley, Charles Colwell,
Lise Villeneuve, R.W. Bradford,
Stephen Payne, Edmund Patten,
Claudette St-Hilaire, Keith Wilson,
Geoffrey Cooke, A.J. Shortt,
Peter Jessen.
MISSING:
Dorothy Fields, George Kearney,
Greg Dorning, Chris Peach.
(NAM 18262)

and W/C D.P. Wightman (Sopwith Snipe) under the leadership of W/C P.A. Hartman (Sopwith Camel), as well as some ground support crew. Unfortunately the Nieuport 17 suffered a failure in the engine mounting and crashed on a practice flight. F/L Long escaped with only minor injuries but the machine was badly damaged, although it was later repaired by 6 Repair Depot at Trenton. Subsequently the Sopwith Triplane was withdrawn from the flying display as a precautionary measure although it had been flown successfully without problems.

The flying display on June 6 was a success and with this precedent Dr. Baird decided to continue to fly some machines in subsequent years. Museum aircraft have now been flown in all provinces except Newfoundland. Pilots have been P.A. Hartman, RCAF retired, George Neal, and Captains Wayne "Butch" Foster and John Williams of the Canadian Forces.

After the flying display in June the CWM staff at Rockcliffe were removed and put to work to establish the exhibits in their new larger building in Ottawa. This left Rockcliffe short-handed, with the result that the aircraft restoration program ceased with the completion of the Aeronca C-2 in May 1967. I resigned in September 1967 and R.W. Bradford was appointed Curator. In 1969 A.J. Shortt was taken on as Assistant Curator.

At the end of 1981 Dr. Baird resigned as Director of the National

R.W. Bradford, now Associate Director, National Aviation Museum.

Museum of Science and Technology and Bradford was appointed Acting Director, a post which he continued to fill until a new Director, Dr. J.W. McGowan, was appointed on January 1, 1984. During this period Bradford had little time for NAC curatorial duties, but he was in a position to negotiate resources and recognition for the National Aeronautical Collection. In September 1982 the name National Aviation Museum was proposed to and accepted by the Board of Trustees of the National Museums of Canada. It was to operate as a subsidiary museum of the National Museum of Science and Technology but with the intention that it become an independent national museum upon the future amendment of the National Museums Act by Parliament. To further this idea, Bradford was appointed Associate Director of the Museum on January 1, 1984, with complete control of its activities except for some of the highest administrative matters.

The Search for Specimens

Perhaps the most interesting part of the history of the National Aviation Museum is how major specimens were found and obtained and how the Museum grew from a modest beginning to its present stature. This growth has been achieved partly by the Museum staff and partly by the generous assistance of others.

Finding the "Canuck"

From the beginning it was realized that any "national" aviation collection had, somehow, to obtain a Curtiss JN-4 (Can.) or Canuck, a member of the Jenny family. The Canuck has more Canadian "firsts" to its credit than any other single type: It was the first to go into large-scale production in Canada; it did the first military flying in Canada; it flew the first Canadian air mail; it made the first aerial survey in Canada and possibly the first in the world; it made the first flight over the Canadian Rockies and finally, it was the mount of most early Canadian barnstormers beginning in 1919 and thus was the first aircraft seen close-up by many Canadians.

It was also known that this acquisition might be difficult, as the Canuck did not seem to have survived as well as the American Jenny, and even if one was found the Museum had no funds with which to purchase it. Indeed the thinking seemed to be that, as the Museum was small and just starting, it needed little funds. In fact the reverse was true, for it needed everything, and the staff realized that the time for gathering certain specimens was rapidly running out.

Just when the first inquiries were made is not known, but a letter of May 15, 1961, to H.S. Fyfield in Connecticut proved to be the one that

The Curtiss JN-4 (Can.) with Edward T. Faulkner who had it for 37 years. This photo was taken on February 15, 1962, at Honeoye Falls, N.Y., just before loading it for transport to Ottawa.

(NAM/W. MERRIKIN 4373)

led to the acquisition of the Canuck. Fyfield was known to have the components of two JN-4s and one was believed to be a Canuck. He replied that both his aircraft were JN-4Ds but that Edward Faulkner of Honeoye Falls, New York, just south of Rochester, did have a Canuck. A letter was sent to Mr. Faulkner on June 19 but no reply came. This resulted in a quandry, for there was little doubt that Faulkner's machine was in fact a Canuck; this might be a unique chance to obtain a most important specimen. It should be mentioned that members of the Curtiss JN family of aircraft were similar, and many people, even so-called aviation buffs, could not differentiate between them; however, Fyfield could do so.

Later that summer, Dr. Loris Russell, the newly appointed Director of the Museum of Human History, to whom the Museum was then reporting, expressed interest in acquiring a Canuck. This was un-expected, as the subject had not been mentioned to him. He said he had asked the Canadian War Museum to try to obtain one earlier without results. The situation with Mr. Faulkner was then explained to him and it was suggested that we visit Mr. Faulkner. This was done on September 11 and the aircraft was determined to be indeed a Canuck. Canada's special interest in obtaining such a machine was explained and the owner showed a stack of letters about four or five inches thick asking about the aircraft, none of which he had answered. No proposition was pressed on Mr. Faulkner at the time. It was felt that an agreement could not be reached by mail and a return visit was purposely delayed until November 6. At this time he agreed to sell the aircraft for $8,000, a figure that was later raised to $9,000 by correspondence. How Dr. Russell found the funds for this important specimen is not known, but he did so and all Canadians should be eternally grateful to him.

William Merrikin and Arthur Walker were sent to disassemble the aircraft and prepare it for shipment in February 1962. It was known that at least one pair of wings, assembled as a unit, was stored on the rafters, and Merrikin and Walker soon found that they could not be removed. It seems the barn had been moved from its original site across the road with the aircraft inside. On its new site, the old door, which had given clearance to swing the wings up, had been sealed in and a new door had been cut on the opposite side of the barn. So Merrikin and Walker had to disassemble the wings standing on the rafters, swaying in a cold wind that whistled in through the open door that provided the only light for working. The machine was loaded into a furniture van for shipment to Ottawa on February 15.

W/C Manning and the RCAF were very interested in the Canuck and requested that it be displayed at Rockcliffe on Air Force Day, June 9, 1962. It was agreed to show it stripped of fabric and its display in this form attracted much favourable comment. Years later in 1983, W/C Manning remarked that he felt the acquisition of the Canuck was what had sparked the interest of the RCAF and the Canadian War Museum in building up their historic aircraft collections. The acquisition of the Curtiss JN-4 (Can.) was a special event because of its great historic importance, but specimens were badly needed in the categories of World War I, bush, transport, pioneer and general aviation aircraft.

Acquiring the World War I Machines

Canadians played a prominent part in the first war in the air and served in the British air services in numbers out of proportion to the relative population of Canada in the British Empire. It was therefore necessary to obtain a good selection of World War I machines both to commemorate the Canadian contribution in the war and to show the development of the flying machine in that period. This would be difficult as most existing machines were held by institutions, and while some could be obtained with hard work and good luck, reproductions would have to be obtained to round out a suitable display of World War I aircraft.

The Nieuport 17 as originally finished by its builder, Carl Swanson, at Sycamore, Illinois. It was finished as N2474 of N.124, the Escadrille Américaine, *popularly known as the* Escadrille Lafayette. *Its armament, cowling and markings were changed and drag wires added to conform with Bishop's aircraft of No. 60 Squadron, RFC, before delivery to the Museum. It is now known that Bishop's machine also had its flight letter displayed and this will be added.*

(R. STOUFFER)

The Museum has always adhered to the principle that the only reproductions suitable for a national museum are those made to original drawings and fitted with original engines, so the Museum's reproductions differ from the originals only in the year of manufacture and, of course, their maker. Many reproductions flying today do not measure up to these standards. Of the eight World War I aircraft that have been added to the collection since 1960, five are original specimens and three are fine reproductions.

The first of the World War I aircraft obtained was the Nieuport 17 reproduction. The late Frank Tallman, when queried, described it as being of museum quality. Of course, it was desirable to have an artifact associated with Lt. Col. W.A. Bishop, VC, and that was one of the classic fighter aircraft of the war. A personal inspection revealed its construction to be first class – no shortcuts had been taken. However, it had been built as a French machine of the *Escadrille Américaine* N.124 of the French *Aviation Militaire*, later better known as the *Escadrille Lafayette*. Thus it differed in armament, markings and other details from the Nieuport 17s of the RFC, but its builder, Carl Swanson of Sycamore, Illinois, readily agreed to carry out the necessary changes should the Museum decide to take it. There were no funds available to purchase it but fortunately a generous donor, who wished to remain anonymous, was found and it was acquired in the spring of 1963 after being modified to represent Bishop's VC-winning aircraft.

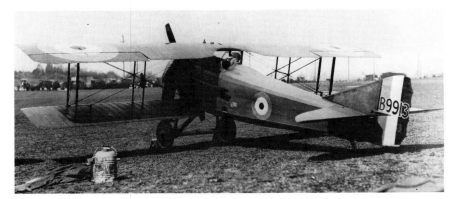

While the Museum was in the throes of obtaining the Nieuport 17, the Sopwith Snipe so carefully restored by Jack Canary in California was offered for sale in *Trade-A-Plane*. The Snipe was the last British fighter to go into service in the war and, although its performance was disappointing, it was immortalized by Major W.G. Barker's VC-winning combat with about 60 enemy aircraft on October 27, 1918, in which, although wounded, he was credited with bringing down four enemy machines. Obviously it was desirable that Canada obtain the Snipe but the Museum had its hands full with the Nieuport so it advised the Canadian War Museum of the availability of the Snipe. The CWM immediately entered into negotiations that were concluded successfully in February 1964. The Snipe was first flown at Rockcliffe on May 21, the first CWM aircraft to be flown while in its possession. (For a history of this interesting machine and details of Jack Canary's fine restoration, see Appendix 2.)

The next World War I aircraft obtained was an original Spad 7. In this case the Museum was informed of the availability of the aircraft by Robert Stuart of Oshawa. It was found to be the Spad restored by James Petty of Gastonia, North Carolina, and on loan to the USAF Museum at Wright-Patterson AFB. Although the Spad 7 and the similar, later Spad 13 are usually thought of as flying with air services other than British, the type was used by the RFC, which had six Spad squadrons on operations and five on training duties. Not only was the Spad 7 one of the classic fighters of the war, but also the highest scoring Spad pilot of the RFC was a Canadian, and it was thought that a Spad 7 would make a fine addition to the World War I specimens. The aircraft was purchased and brought to Rockcliffe in December 1965.

The Sopwith Triplane reproduction was the next World War I aircraft to be added. When the Museum was taking delivery of the Nieuport 17 in 1963, it was remarked to Carl Swanson that if he would make a Sopwith Triplane with the same standard of workmanship, the Museum would be interested in acquiring it and could supply the engine. The Sopwith Triplane was the type used by the famed all-Canadian Black Flight of No.10 (Naval) Squadron, and it was thought that an example of the now-rare triplane configuration should be held by the Museum. As original drawings were available, a first-class reproduction could be made. Swanson proceeded with construction in his usual meticulous way and the Museum took delivery of the Sopwith Triplane in the spring of 1966.

The next World War I type acquired was the Avro 504K, the famed RFC/RAF trainer and later the standard military trainer of the RCAF until 1928. This acquisition came about because the RCAF wanted to fly an old RCAF type in the display they planned for the Centennial Year celebrations in 1967. By good fortune they were able to arrange the purchase of an original 504K from Cole Palen of Rhinebeck, New York. They then bought a second 504K, which had been partially restored with all new wood, from Major J. Appleby in California. It was then decided to keep the Palen aircraft as a Museum specimen and to use the Appleby machine and a completely new RCAF-made reproduction in the flying displays. All three specimens were turned over to the National Aeronautical Collection in 1967 following the RCAF flying displays.

The Fokker D.VII was the first enemy aircraft type acquired. The late Dave Fox told Bob Bradford that the original Fokker D.VII held by J. Nissen and the late J. Mathiesen in California would be put up for

A Hall-Scott L-6 powered two-seat Fokker D. VII at Oakland, California, during the filming of Hell's Angels *with pilot Ross Cook in the cockpit. The aircraft were not individually identified in film use, which makes it difficult to distinguish between them; the numeral shown here is an exception. It was believed to be the Museum's specimen, but now it seems this may not be so.*

(VIA H.H. WYNNE)

sale. The type was the leading German fighter in 1918. Only seven are known to have survived and this specimen was the only one in private hands. It had been used in Hollywood movies of the late 1920s and early 1930s. After inspection by Bradford, it was bought and shipped to Rockcliffe in February 1971. As described in the restoration section of this book, its restoration is well advanced but not complete at the time of writing.

The Sopwith Pup reproduction built by George Neal at Toronto was purchased in October 1973. What is noteworthy about this fine reproduction is that every bit of its structure was made personally by its builder, and the only assistance he received was in the application of its fabric covering. Neal was its pilot for all of its early flying and for most of the flying it has done since its acquisition.

The last World War I aircraft obtained to date is the superb original Maurice Farman S.11 Shorthorn bought at the Wings and Wheels Museum auction at Orlando, Florida, in December 1981. The specimen is one of the few surviving examples of the pusher-type aircraft that played such a prominent role in the Allied air services at the beginning of the war. The need for this specimen to fill a conspicuous gap in the collection had been realized for some time and an unsuccessful bid was made for it at the 1968 auction of the Tallmantz Aviation Inc. collection. Thirteen and a half years later the Curator's plea for more funds to compete at the Orlando auction was approved by the Director and Board of Trustees and a successful bid was made.

The Addition of Bush Aircraft

The next class of aircraft that ought to be well represented in the
Museum's collection was the bush aircraft that had contributed so
much to Canada's development by opening up northern areas in the
1920s and 1930s. In 1960 the only example preserved was the Noorduyn
Norseman VI in the RCAF's collection.

High on the list of desirable bush aircraft was the Fairchild FC-2W-2,
or its almost identical successor, the Fairchild 71. They were the most
numerous of the types that formed the Canadian bush flying fleet from
1928 through the 1930s and did the lion's share of the work. On my first
visit to the National Air Museum in Washington in 1961 I asked Head
Curator Paul Garber if he knew of a specimen of either of these
Fairchild models that might be available. Paul replied that one had been
offered to them not long before but was turned down as it was similar
to the Pan American Airways' Fairchild FC-2 they already had. The
aircraft in question was an FC-2W-2 model owned by Virgil Kauffman,
President of Aerial Surveys Inc. of Philadelphia. Subsequent correspondence with Mr. Kauffman established that the machine had been
damaged in a forced landing but was repairable and was without
engine and propeller. Mr. Kauffman generously agreed to donate the
aircraft to the Museum and it was brought to Ottawa in June 1962. This
most helpful gift was especially appreciated because of its American
origin, and its restoration was started immediately after the completion of the Curtiss JN-4 (Can.).

In the meantime, late in 1961, it was reported that the Junkers W.34
CF-AQB was available in British Columbia. W.J. Jacquot, an air engineer with experience on Junkers machines, was asked to inspect the
aircraft. He reported that it was in poor condition and would be

difficult to restore even to display condition and suggested that we consider CF-ATF, another W.34 that was also available. This was the machine that had been offered to the Museum Committee in 1959 and turned down. It had not been mentioned to the Museum on its formation or it is likely that earlier action would have been taken.

The owner of CF-ATF, Pacific Wings Ltd. of Vancouver, said that it was available as is or for a slightly increased cost if they were to put it in ferriable condition. It was wonderful news to hear that this specimen from the golden age of bush flying was not only available but ferriable. Its interesting structure and fine contribution to aviation throughout the world made it doubly valuable. The difficulty was that the NAM had no money for such a purchase. This troubling situation continued for several months with no apparent solution and with the worrying thought that this historic Canadian aircraft might well end its days on the shore of some northern lake freighting fish.

Finally, and suddenly, the happy thought occurred that Mrs. J.A. Richardson might well be interested in preserving CF-ATF. She was the widow of James A. Richardson, President of Canadian Airways Ltd., the original owner of CF-ATF and a man who had done more than anyone to develop Canadian civil aviation in the late 1920s and the 1930s. Indeed so much had he done that he became known as the "Father" of Canadian aviation. Following an exchange of letters, Mrs. Richardson agreed to donate the aircraft to the NAM at the end of July. It seems likely that this fine historic aircraft would not have survived without her thoughtful and generous action.

Pacific Wings selected TCA First Officer R. John Racey as the ferry pilot, and CT-ATF left Vancouver on September 8 for Prince Albert, Saskatchewan, via Kamloops, British Columbia, and Cooking Lake, Alberta. On the 9th it went on to Winnipeg via Yorr Lake. The aircraft was left there until September 15, when it went on to Sault Ste. Marie, Ontario, via Kenora and Port Arthur. On the 16th it flew to Pembroke, and on the 17th made the brief flight to Rockcliffe. Mrs. William Benidickson presented it to the NAM on behalf of her mother, Mrs. Richardson. A total of 21 hours 5 minutes flying time was taken for the trip and the total time logged on CF-ATF during its 30-year active lifetime was 12,209 hours.

CF-ATF was stored over the winter, and in the spring, with the help of the RCAF, it was cleaned and repainted in the markings of Canadian Airways and placed on display. Over its long Canadian Airways career it had worn a number of slightly different markings. To be typical of the well-known Junkers markings in Canada, it was given the orange wing and fuselage bands, the Junkers serial number on the fin, and the famous Flying Goose insignia of Canadian Airways. Although no photograph was found with CF-ATF wearing all three of these markings together, it was decided to take poetic licence and apply them.

The Museum's Bellanca Pacemaker in service in Alaska with its previous owners, Alaska-Coastal-Ellis Airlines.
(NAM 4618)

There have been no complaints, and one suspects complaints might have been received had this not been done. CF-ATF is one of two W.34s surviving in museums, the other being in Sweden.

In 1963 in *Trade-A-Plane*, a Bellanca CH-300 Pacemaker was offered for sale in Juneau, Alaska. This was of great interest, not only as a fine bush aircraft but also because it was used on many long-distance flights due to its weight-lifting capability. Its owner, Alaska-Coastal-Ellis Airlines, Inc., seemed pleased that the machine was wanted for preservation and their price was reasonable. Again, the difficulty was the Museum had no money. Time dragged on and schemes were thought of to raise money or interest a possible donor, but none of these worked out, and I feared that it might be acquired by someone else.

Fortunately, early in January 1964 and almost a year since the Pacemaker had been advertised, the Museum was advised that funds for its purchase would be forthcoming. The machine was still available and we arranged to pick it up in May and fly it to Rockcliffe on floats. Our ferry pilot was A.G.K. (Gath) Edward, Assistant Director of Flight Operations of TCA and a former Pacemaker pilot in northern Quebec in the 1930s. I went along as crewman. Gath, a good friend of the Museum, had followed our efforts to obtain the Pacemaker with keen interest and had even worked out a scheme for its donation but without success.

We arrived in Juneau in the morning of May 26 and, to take advantage of fine weather across the continent, took off late in the afternoon for Prince Rupert, British Columbia. Next morning a scenic flight up the Skeena River took us to Prince George but an intended flight to Cooking Lake had to be terminated at McBride, British Columbia, owing to local weather conditions. An attempt to get away next morning was again thwarted by weather, so the flight to Cooking Lake was only completed in the afternoon. The next day we flew to Sioux Lookout, Ontario, via Prince Albert, The Pas, and Lac du Bonnet. The following day, May 30, we flew to Rockcliffe via Nakina and Sudbury, landing on the Ottawa River at 7:00 p.m. This is believed to be the last flight of any member of the fine family of Bellanca aircraft which made so many long-distance flights a half century ago.

In 1964, the Museum carried out its first retrieval program to bring back the remains of another bush aircraft. The remains of a Fleet 50, CF-BXP, were reported on the shore of Sandgirt Lake, Labrador (now swallowed up by the Smallwood Reservoir). As it was an interesting bush aircraft of Canadian design, the Museum decided to salvage it. Two of the Museum staff, William Merrikin and Omar Clarke, flew in to disassemble the components and prepare them for airlift out by RCAF helicopter to the nearest road. In 1968 M.L. McIntyre salvaged a wing panel and other smaller components of another Fleet 50, CF-BJW, from O'Sullivan Lake, Ontario, northwest of Nakina, to aid in its restoration. Restoration has not been started and it seems likely that it will be deferred until higher priority restoration programs – and there are several – have been completed.

The Fairchild 82A, CF-AXL, was donated to the Museum in May 1967 during a dinner at Ottawa to celebrate the 25th anniversary of the formation of Canadian Pacific Airlines. The Museum had had an opportunity to purchase it earlier but had turned it down with regret owing to the usual lack of funds. However, it was shortly offered to CPA and arranging its purchase was one of the last acts of G.W.G. McConachie, President of CPA, before his sudden death in June 1965. The aircraft had been originally owned by Starratt Airways and Transportation Ltd., which was acquired by the Canadian Pacific Railway, and it and nine other companies bought about the same time formed the nucleus around which CPA was created. The aircraft was finished in CPA's early markings before being presented to the Museum.

The Museum also needed specimens of the flying boats that had predominated in northern flying in the 1920s. While the Curtiss HS-2L was the most important, there were other types of considerable historical interest. It was known that the Science Museum at South Kensington in London held a Curtiss Seagull. It was also believed that there was little British interest in the type as no Seagull had ever flown in Britain. On the other hand, the type had been used in Canada and would be a significant relic of our early days, especially as it seemed unlikely the Museum would ever have an HS-2L.

In the summer of 1965 I set off on what was ostensibly a holiday

with my wife in England but what was actually planned as an effort to obtain the Seagull for the Museum. One can imagine my surprise when, as I enjoyed a cup of tea with the staff of the Science Museum, William T. O'Dea, Keeper of the Department of Aeronautics and Sailing Ships, suggested to Brian Lacey of the Aeronautics Department that the Seagull be given to Canada. This remarkable turn of events confirmed our thinking on the relative importance of the Seagull to the two museums. I immediately said how much the aircraft would be appreciated in Canada and that the Museum would be very pleased if it could be obtained.

In order to make the transfer to Canada, the Science Museum had to make sure no other British museum was interested in the Seagull. This took some time so when Dr. Baird visited England in the fall of 1967 he met with Sir David Follett, Director of the Science Museum, and arranged that the matter be expedited. The Seagull arrived at Rockcliffe in May 1968 on what was initially agreed to be a long-term loan. It was soon determined that an extensive restoration would be required before the aircraft would be in good display condition, and, understandably, the Museum (now renamed the National Aeronautical Collection as part of the NMST) was reluctant to put time and money into the specimen if it was only to be a temporary visitor. It was then agreed that the Seagull would be given to the Museum in exchange for the nose section of a Douglas DC-3. Restoration started in 1969.

The Museum's Curtiss Seagull during its much publicized survey of the Parima River, a headwater of the Amazon River, 1924-25.
(NAM 14088)

53

It had long been realized that an example of the de Havilland D.H.C.2 Beaver should be obtained, as it had been a highly successful aircraft and was the most numerous of all Canadian-designed aircraft with many still in active use in Canada and elsewhere. Most desirable of all from the Museum's point of view was the prototype, CF-FHB, which was first flown in August 1947. So when CF-FHB was offered for sale early in 1979, negotiations were started with its owners, Norcanair Ltd. of Prince Albert, Saskatchewan. The difficulty was that Beavers were, and still are, much in demand even after over 30 years and the Museum could only afford about 70 per cent of the asking price. The owners agreed to remove it from sale while the Museum tried to work something out. Dr. Baird encouraged de Havilland of Canada to make a donation, and, late in 1979, the Molson Foundation generously agreed to make up the balance. CF-FHB landed on the Ottawa River at Rockcliffe on floats on July 29, 1980.

The Stinson SR Reliant was acquired by purchase in 1983 and flown on floats to Rockcliffe. The type had little Canadian history, with only one, CF-AUS, being used in Canada before World War II; however, it marked the beginning of the Reliant series of aircraft and is the only SR Reliant preserved by any museum at present. The Museum's specimen had been imported in 1953.

The Curtiss HS-2L is probably the most interesting of all the Museum's accessions, as it has necessitated locating and obtaining parts and components of three different aircraft and it represents the longest and most difficult restoration program yet undertaken. A Curtiss HS-2L flying boat, the first Canadian bushplane and the type that established the traditions of Canadian bush flying, was high on the list of desirable specimens from the start, but prospects of finding one did not seem good. None were known to exist; flying boats, as a class of aircraft, have not survived well and the large size of the HS-2L would make its storage difficult.

The first indication that we might be able to assemble an HS-2L came in 1965 when the late Frank Tallman, in response to a query, said that the Los Angeles County Museum had one. The L.A. Museum said, however, they had had an HS-2L but the hull had been stored outside and had deteriorated and been scrapped. The wings were in storage but not accessible. An active attempt to obtain the wings was not made at the time because of their inaccessibility and because it was felt more parts would be required before the reconstruction of a complete machine would be feasible.

Next, Donald Campbell of Kapuskasing, Ontario, reported in 1967 that he thought there was the wreck of an HS-2L in a lake nearby. He would try to find its exact location the next year and would report the results. A quick check of available records indicated that it was probably G-CAAC, the first aircraft of Laurentide Air Service Ltd., our first bush flying company, and consequently the very first Canadian bush aircraft. As records were incomplete and the Ontario Provincial Air Service had also been active in the area with their HS-2Ls from 1924 onwards, the identity was by no means certain. In the summer of 1968 Don Campbell located the wreck in a small unnamed lake and reported it in September. The lake was about a mile west of the Groundhog River and ten miles northeast of Remi Lake, a base for Laurentide and later the OPAS. It was named Foss Lake in 1970 after G-CAAC's pilot, D.B. Foss.

Curator R.W. Bradford flew into the lake with Don Campbell in September, salvaged a few small items and was joined by two of the Museum's staff at the end of the month. More small items were retrieved before the recovery work had to be abandoned with the worsening October weather. In the meantime, 'AC's pilot was located and he confirmed the location as that of 'AC's crash.

With the cooperation of the Honourable René Brunelle, Ontario's Minister of Lands and Forests, plans were laid for salvage operations in the summer of 1969. A salvage party made up of Museum personnel, a volunteer diving team and members of No.647 Air Cadet Squadron from Kapuskasing were flown to the site. A raft was constructed and the remains of the hull and other components were raised and brought to shore. A large amount of silt which had preserved much of the hull, had to be removed by pump before it could be raised. Building the raft, salvaging the hull, searching for small parts on the lake bottom and treating wood parts with preservative went on between August 7 and September 12, and the hull was airlifted out on the last day. The hull still bore the registration G-CAAC, confirming it was Canada's first bush aircraft, which had been named in 1919 *La Vigilance*.

After these encouraging results, Bradford inquired into the availability of HS-2L drawings. Fortunately they had survived. Capt. L.C. Powell, USN, kindly arranged for the donation of the drawings which proved to be an almost complete set. The Museum followed up on the surviving wings in California and found that the complete tail surfaces and a complete set of struts and wires had also survived there. The Los Angeles County Museum did not have a clear title to them, but after prolonged negotiations the owner agreed to accept a 130 hp Clerget rotary engine in exchange for the components. These components had belonged to an HS-2L of Pacific Marine Airways which operated a passenger service between Wilmington, California, and Catalina Island off the coast during the 1920s. During this time, donations of HS-2L instruments and a propeller were made by helpful individuals.

It was decided that the remains of 'AC's hull would be retained as a historic exhibit, less some metal fittings which would be installed in a new hull started in 1975. Meanwhile, the remains of another Canadian HS-2L, G-CAOS of the OPAS, were reported to the Western Canada

LEFT
Laurentide Air Service's Curtiss HS-2L, G-CAAC, at Remi Lake, Ontario, shortly before its accident. It crashed at a lake now known as Foss Lake.

(DND RE-64-3183)

RIGHT
G-CAAC after its accident on September 2, 1922.

(NAM 2749)

FAR LEFT
An aerial view of the Museum's camp in the bush at Foss Lake and the salvage raft.

(NAM)

LEFT
A painting by R.W. Bradford illustrating the methods employed to salvage G-CAAC. Two divers underwater direct suction tubes into the hull to clear the accumulated mud while the crew above tends the pumps and looks after the needs of the divers.

(KMM)

TOP RIGHT
The largest single portion of G-CAAC salvaged was photographed with the salvage crew.

(NAM)

BOTTOM RIGHT
A number of components, including the wings and tail surfaces, of this Curtiss HS-2L were located in California and obtained by good fortune. They made a valuable contribution to the restoration.

(WESTERN AIRLINES)

Aviation Museum in Winnipeg. This aircraft had been forced by engine failure to land on the quite unsuitable Kenogami River near Longlac, Ontario, about 340 km west of 'AC's crash site. In the summer of 1978 a joint retrieval operation was carried out by the two museums. All wooden parts were found to have rotted away but the engine, radiator and metal fittings were recovered. The Liberty engine went to the Western Canada Aviation Museum as agreed, while the radiator and fittings were retained by the NAC. By good fortune the fittings nicely complemented those already obtained from 'AC. The radiator was the last large item needed for the restoration. The HS-2L now on display at Rockcliffe incorporates parts from three different machines and is the result of cooperation by several organizations, good luck, diligent detective work and, finally, very skilful work by the Museum staff.

Donation of Transport Aircraft

While many aircraft have been donated to the Museum, it is only in the transport aircraft category that every specimen has been donated. Indeed if they had not been given, it is unlikely they would ever have been acquired as funds have never been sufficient to afford acquisitions of their value. Canadians should be grateful to the several organizations that made the preservation of this portion of our aviation history possible.

In 1962 it was heard that the veteran Lockheed 12A light transport CF-CCT was to be retired by the Department of Transport the following year. The DOT agreed that, in view of its pioneering airway development work and its "dawn-to-dusk" Montreal-Vancouver survey flight of 1937, the aircraft should be placed in the Museum. CF-CCT was meticulously refinished under the supervision of its former co-pilot/pilot, J.D. Hunter, before being presented to the Museum on September 24, 1963.

The next to arrive was a military transport, the Douglas C-54GM North Star which had been flown in on September 28, 1964. This was, of course, only a transfer of the aircraft by the RCAF to its historic aircraft collection, but its arrival has a curious story attached to it. In 1963 W/C Manning and I discussed the impending retirement of the North Star and agreed that one would not be requested for several reasons, one of which was that it was too large to be accommodated at Rockcliffe. Soon after the matter of the North Star retirement was raised in Parliament, where a member more or less insisted that one be sent to Rockcliffe. Since its arrival it has had to be stored outside and has inevitably suffered.

In December 1966 the Chief Pilot of Chevron Oil Co. in Calgary phoned to say that his company had a Boeing 247 that would be retired

shortly and asked if the Museum would be interested in having it. Now the Boeing 247, the first of the modern airliners, was high on the list of wanted specimens and this was the first time a generous and obviously thoughtful person had ever come forward on his own initiative and proposed such a fine donation. There was probably at least a moment's silence before I could recover my equilibrium enough to thank Mr. Martin for this thoughtfulness and to say that the Museum would be delighted to have it. The presentation was cleared with the company's American parent, Standard Oil of California, and the aircraft was flown to the Museum at Rockcliffe on a bitterly cold February 11, 1967, and turned over. It is one of four of this historic type preserved in museums, one each in Britain and Canada and two in the United States.

Early in 1968 Air Canada asked if the Museum would like to have the Lockheed 10A Electra CF-TCA. The offer was gladly accepted.

The Museum's Boeing 247 taxis in at Rockcliffe on a bitterly cold and windy February 11, 1967. The Boeing was the first of the modern airliners and this specimen was a particularly valuable accession.

(NMC J19200-4)

59

CF-TCA was not TCA's first aircraft but was the first received directly from its manufacturer. After only a brief TCA service, it went to the RCAF and then to the United States. In 1967 Mrs. Anne Pellegreno and crew flew it around the world following Amelia Earhart's route of 1937. It was bought by Air Canada in March 1968 and restored by the airline to its 1937 condition. H.W. Seagrim, Air Canada's Executive Vice-President, presented it to the Museum at Ottawa International Airport on October 15, 1968, and it was accepted by Dr. D.M. Baird. It was displayed at the Museum Annex at the airport until the 1970s, when it was placed in storage at Rockcliffe. It was displayed at Expo 86 in Vancouver in 1986.

The idea of presenting a Vickers Viscount on the type's retirement from Air Canada service had been discussed as early as 1967 and it was agreed that it would be a highly desirable addition to the transport aircraft section. The Viscount went into service on TCA's routes in April 1955, the first turbine service on the North American continent. Viscount CF-THI was refinished in its original TCA markings and flown to Rockcliffe on November 19, 1969, where it was presented to the Museum by Air Canada. CF-THI was the first propeller turbine-powered aircraft to join the collection.

It had long been felt that a specimen of the classic and apparently immortal Douglas DC-3 was an essential addition to the transport aircraft collection. It had to be a real DC-3 and not the military conversion, the C-47, and it had to have a long and interesting Canadian history. In January 1983 an article appeared about the Goodyear Tire and Rubber Company's DC-3, which was due to be retired shortly. It seemed the most suitable DC-3 for the Museum and further checking only served to confirm the first impression. The aircraft had been on order with Douglas as a DC-3 when the U.S.A. entered World War II and it was impressed into the USAAC. In 1945 it became TCA's first DC-3, and in 1948 it was bought by Goodyear and served as their company aircraft, based at Toronto, until 1983 – 35 years with the one firm. It is unlikely that any other DC-3 has exceeded this service. The Goodyear flight crew agreed that it would be desirable that the aircraft, C-FTDJ (earlier CF-TDJ), go to the Museum and Curator R.W. Bradford concurred that there could be no finer or more suitable example of this historic type for our national museum. The company generously agreed to donate the aircraft and it was flown to Rockcliffe and handed over to the Museum on December 19, 1983.

Although many World War II military aircraft had been preserved by the RCAF, some were missing, and there were no RCAF aircraft of the between-the-wars period. While a few more wartime aircraft have been added, the machines of the between-the-wars period have proved difficult to find. As this is written, types such as the Armstrong Whitworth Siskin and Atlas, Westland Wapiti and Canadian Vickers Vedette are all conspicuous by their absence, and with the possible exception of the Vedette it seems unlikely these gaps will be filled.

The Museum first found that Bancroft Industries of Montreal owned some surplus RCN Hawker Sea Furies in 1963 and started to take steps to obtain one. The Museum was persuaded to let the Canadian War Museum pursue this type and Bancroft donated one that year. It was restored by the RCN and Fairey Aviation of Canada Ltd. before being brought to Rockcliffe.

In 1946 George A. Maude of Sidney, British Columbia, purchased a war-surplus Bristol Bolingbroke IVT and moved it to his property on Saltspring Island off the east coast of Vancouver Island. In 1962 in correspondence with W/C Manning, Maude learned that the RCAF lacked a specimen of the Bolingbroke and agreed to donate his. The RCAF sent a crash retrieval team to look at the machine with a view to taking it off the island; they reported it could not be done. As two men with no equipment had taken it there they were asked to take another look. In the event the RCAF had CPA (Repairs) Ltd. of Calgary disassemble the Bolingbroke and truck it to Calgary for restoration in May 1963. CPA restored it to excellent display condition and it was finished in the markings of an aircraft of No.8 Squadron, RCAF, by February 1964. The Bolingbroke was then transported to Rockcliffe and officially handed over by George Maude to the RCAF during the Air Force Day display on June 6, 1964.

In 1964 Dr. Alexander D. McLean, a Toronto dentist and member of No.400 RCAF Reserve Squadron, was in touch with W/C Manning over a project to restore a Supermarine Spitfire to flying condition with the intention of inspiring Canadian youth. The Museum learned from Carl Swanson of Sycamore, Illinois, the builder of their Nieuport 17 and Sopwith Triplane, that there was a former RCAF Fleet 16B Finch II for sale at Sycamore. This information was passed on to W/C Manning, who apparently suggested to Dr. McLean that restoring the Fleet would be easier than the Spitfire project and would fill an outstanding gap in the collection of World War II aircraft. Dr. McLean acted promptly and bought the aircraft in early September. The restoration work was undertaken by No.400 Reserve Squadron at Downsview. The Finch was finished in its original RCAF markings with civil registration CF-SUX. Its first flight after restoration took place on July 1, 1966.

John Paterson's Supermarine Spitfire IX L.F. in its 1962 Canadian civil markings but with the Red Indian insignia of his old squadron, No. 421, RCAF, added to the nose.
(KMM/H. TATE)

John Paterson (right) hands over his Spitfire IX L.F. on Air Force Day at Rockcliffe, June 6, 1964. K.M. Molson accepted for the Museum.
(DND PL 144612)

In 1961 John N. Paterson of Fort William contacted the Museum to see if it could supply technical information he required to restore his Supermarine Spitfire IX. Unfortunately the information could not be supplied, but the Museum expressed interest in his machine should he ever decide to retire it. He decided to do so after a flying demonstration to a reunion of World War II fighter pilots at Montreal in the fall of 1964. The aircraft was stored at the RCAF Mountain View airfield over the winter, Paterson making his last flight in his Spitfire to deliver it there in October. In the spring of 1964 Jerry Billings flew the Spitfire to

The Museum's Consolidated B-24L Liberator in New Delhi, still in its Indian Air Force markings.
(KMM/H. TATE)

Ottawa, and he flew it again for a last demonstration on June 6, 1964, before Paterson presented it to the Museum. The following day Billings flew it to Ottawa International Airport, where the engine was inhibited (prepared for storage) and the aircraft went on display in the Museum Annex.

In the early postwar years an eccentric farmer, Ernest V. Simmons of Tillsonburg, Ontario, bought a large amount of surplus war material, including about eight Fairey Swordfish and 36 North American NA-64 Yale aircraft. These were stored on his farm and jealously guarded by an unfriendly dog and a shotgun-armed Simmons. The Canadian War Museum was able to buy a Swordfish from him on March 22, 1965, and the aircraft was restored by Fairey Aviation of Canada Ltd. at Eastern Passage, Nova Scotia, and the RCN. After Simmons's death in 1965, the rest of the material was sold at auction on September 5.

In October 1967, a Hispano HA-1112-M1L (the Spanish-built version of the Messerschmitt Bf 109) arrived at Rockcliffe after being purchased in England for the Canadian War Museum. The Spanish-built Bf 109s were made in the postwar period and fitted with Hispano-Suiza or Rolls-Royce Merlin engines in place of the Daimler-Benz DB 605 engines used by the Germans.

The acquisition of the Consolidated B-24L Liberator is an interesting story. In 1967 Air Chief Marshal Arjan Singh, Chief of the Air Staff of the Indian Air Force, was visiting Canada and visited the National Aeronautical Collection at Rockcliffe with Air Marshal E.M. Reyno.

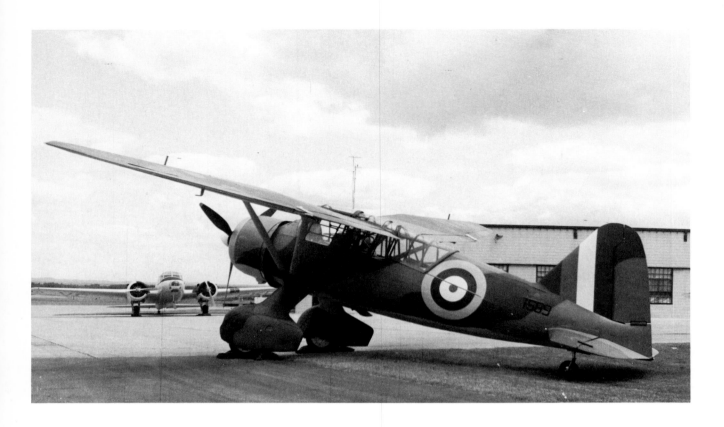

The Westland Lysander III, RCAF 1589, at Rockcliffe prior to being donated to India. The gift resulted in the unexpected donation to Canada of a Consolidated B-24L Liberator.

(KMM)

The two men had attended the Imperial Defence College together and knew each other well. On seeing the Westland Lysander III, A/C/M Singh became most enthusiastic. The type had been used by the IAF in its early days with both its No.1 and 2 Squadrons and later by No.4 Squadron and, of course, he knew it well. A/M Reyno checked into the Canadian Lysander situation and found a second Lysander was nearing completion at Winnipeg as a Centennial project of 402 (Auxiliary) Squadron. He arranged for the Rockcliffe Lysander III, RCAF 1589, to be sent to India as a gift, and it was flown to New Delhi in RCAF Lockheed Hercules 10321 over September 5-7, 1967. It was promptly assembled and put on display in the IAF Museum in Palaam, New Delhi, after an impressive hand-over ceremony.

At this time A/C/M Singh learned that Canada did not have a Liberator in its collection. The type was just then being retired from the IAF, so Singh arranged for the gift of a Liberator, HE773, to Canada. Col. A.J. Pudsey and Maj. John Ratcliffe were assigned to check the feasibility of ferrying the aircraft to Canada. The main difficulty was arranging the necessary visas and overflight clearances, but this was

done and Col. Pudsey and a crew of five men arrived at New Delhi on May 26, 1968, to ferry the aircraft to Canada.

A/C/M Singh handed over HE773 to the Canadian High Commissioner, James George, on May 27. Conversion training was started by the crew and check-outs on the B-24 were carried out from May 28 to June 5. On June 6 HE773 left New Delhi for Canada, and the long trip proceeded as follows: June 6, to Jamnagar, India; June 7, to Bahrain on the Persian Gulf; June 8, to Nicosia, Cyprus; June 9, to Athens; June 10, layover at Athens; June 11, to Decimomannu, Sardinia; June 12, to Lahr, West Germany; June 13, layover; June 14, to Prestwick, Scotland; June 15, to Keflavik, Iceland; and an attempt to reach Goose Bay, Labrador, abandoned; June 16, to Goose Bay, Labrador; June 17, to Trenton. Some 16 900 km had been covered in 69 flying hours, the longest ferry flight of any Museum specimen, and the crew had been up each day at 4.30 a.m. On the Atlantic crossing HE773 had been accompanied by a Canadian Forces Canadair Argus and, except for minor engine problems on the 15th and 16th, the flight proceeded without incident.*

The Liberator crew on its long ferry flight from India, left to right Capt. A.R. Dickson, 2nd pilot; Col. A.J. Pudsey, a/c Capt.; Capt. R.T. Brown, navigator; CWO J.C. Trethowan, flight engineer; MWO H. Tate, servicing chief; Capt. A.R. Woodman, radio officer.

(DND OA 68-218-4B)

* For a detailed account of this interesting trip see "Operation Longhaul" by Col. A.J. Pudsey in *Sentinel*, October 1968, pp.24-31.

Hawker Hind in Kabul, Afghanistan, October, 1975.

(NAM 13420)

A Between-the-Wars Acquisition

Military aircraft of the between-the-wars period have always been sought but the only one obtained to date is the Hawker Hind. The Hind was one of a family of seven close relatives and an eighth not-so-close relative, the Hector, which started with the flying of the Hart prototype in 1928. None of the family became numerous in Canada, although three Harts and six Audaxes were on the strength of the RCAF and four Hinds were used for ground training in 1942-43. When it was discovered in 1974 that some Hinds still existed in Afghanistan, the Museum decided to inquire about obtaining a specimen. The RAF Museum, which had already obtained a Hind from Afghanistan, was most helpful in the early stages. Dr. Baird approached Afghanistan through diplomatic channels and, with the help of the Canadian ambassador, Dr. Baird and R.W. Bradford arranged to visit Kabul to see the aircraft and negotiate with the Chief of Civil Aviation and the Chief of the Afghanistan Air Force. On the recommendation of these officials, President Daoud of Afghanistan agreed to donate the aircraft to Canada to further good relations between the two countries. A.J. Shortt and W. Merrikin left Ottawa on October 23, 1975, for Kabul, where they disassembled the Hind along with another partial airframe, and prepared the components for shipment, all, of course, in the heat of Kabul. The party returned to Ottawa on November 7 and the components were picked up by the Canadian Forces flight and brought to Ottawa. George Neal started restoration of the Hind at Toronto in 1984 and it is expected to be completed late in 1988.

Addition of Pioneer Aircraft

On the formation of the National Aviation Museum in 1960, there were no examples of pioneer aircraft or general aviation aircraft preserved by any museum in Canada. The early aviation pioneers, of course, were those that started the whole industry, and their aircraft were frail and few in numbers, so inevitably still fewer survived. The machines of the general aviation industry – the trainer, charter aircraft, private owner aircraft, for example – are the "grass roots" of the aviation industry and have been a most important part of its history, but, again, none were being preserved in Canada in 1960.

Only one Canadian-built pioneer aircraft, the McDowall monoplane, is known to have survived. It was bought by K.S. Hopkinson in 1958 and, until 1967, it hung from the rafters of his hangar at Goderich, Ontario. Hopkinson died in a flying accident on March 26, 1964, induced, it is believed, by a heart attack. When his estate later put the aircraft up for sale, funds were not available for its purchase. By great good fortune, however, funds became suddenly and unexpectedly available in March 1967 and the McDowall aircraft was immediately purchased. In this completely fortuitous manner Canada's only pioneer aircraft came to rest in the National Aviation Museum.

The only other pioneer aircraft that has been acquired is the Blériot XI which was built in 1911 in California. In this case R.W. Bradford saw the machine when inspecting the Fokker D.VII for purchase. He recommended acquisition of the Blériot and it was transported to Rockcliffe late in 1974. It had been partially restored by its former owners but no further work has been done on it to date.

D.H. "Moth" at Elliott Airport — Hamilton 1928

General Aviation Machines

The D.H. 60 Moth, G-CAUA, 1928. The specimen will probably be refinished in these International Airways Ltd. markings.

(VIA H. SLAGHT)

From the beginning the National Aviation Museum had sought an example of the D.H.60 Moth because of its importance in Canadian aviation in the late 1920s and early 1930s. Moths seemed surprisingly elusive considering their former numbers and it was only early in 1962 that the late D.B. Simmons of Moncton mentioned that C.F. Burke had one in Charlottetown, Prince Edward Island. Carl Burke generously agreed to give the Moth to the Museum. It proved to be an early Canadian Moth, the seventeenth delivered by de Havilland of Canada, which made it that much more interesting. As the Museum was heavily engaged in the restoration of the Curtiss JN-4 (Can.), de Havilland Canada helpfully agreed to supply the material and facilities for its restoration and ten DHC old-timers offered their time to do the work. The restored aircraft was completed in April 1963 and turned over to the Museum in a ceremony at DHC's plant at Downsview on May 10, 1963.

Also from the beginning, a specimen of the Aeronca C-2 was wanted. This machine and its two-seat counterpart, the C-3, sparked the American light plane movement, and in addition its unusual con-

68

The Aeronca C-2 wearing its American civil registration prior to its acquisition by the Museum.
(E.H. SOMMERICH/P.M. BOWERS)

figuration would be in marked contrast to other specimens in the collection. An example was offered for sale in *Trade-A-Plane* in January 1967 and was bought for a very reasonable price in February. It was restored and flown that spring.

In 1968 an Avro 616 Avian IVM was donated to the Museum by the estate of Charles Graffo of Winnipeg. The machine, a competitor of the D.H.60 Moth, needs restoration and consequently has yet to be displayed. Also in 1968, a Travel Air 2000 was purchased from Mrs. K.S. Hopkinson of Goderich, Ontario, with funds turned over to the Museum by the Canadian Aeronautics and Space Institute. The funds had been surplus to the requirements of the Coordinating Council for the 50th Anniversary of Powered Flight in Canada. The Travel Air 2000 represents a class of aircraft that dominated the American general aviation scene in the late 1920s and was widely used in Canada. The Travel Air 2000 is complete but needs restoration and has not been displayed.

Also high on the list of needed specimens was an open cockpit mailplane. These machines represent an era now long gone when pilots donned helmet and goggles and a parachute and flew the mail

The Museum's Stearman 4EM Junior Speedmail was originally registered to Standard Oil of California as a Model 4E, NC784H as shown here. It, and its two sisterships, were generally regarded as the most outstanding members of an elegant aircraft family. Later it served as a workhorse for a crop duster and lost some of its elegance.

(J.W. UNDERWOOD)

without the benefit of radio, good weather forecasts and other modern amenities. For several reasons a specimen of the Stearman 4EM was thought most suitable. On June 6, 1964, after donating his Spitfire IX, John Paterson asked what next he could do for the Museum. Obtaining a Stearman 4E being used as a duster in the United States and restoring it to its former mailplane configuration was immediately suggested.

In *Trade-A-Plane* of January 2, 1965, a Stearman 4E was advertised for sale and John Paterson was immediately advised. He bought the machine on January 28 and it was flown to Fort William from Idaho on February 5 by its former owner. A number of useful Stearman components as well as a Pratt & Whitney Wasp SC engine with which the type had been originally equipped were obtained from an operator of several Stearman 4Es being used as dusters in Fresno, California. The restoration of the aircraft in Fort William proceeded slowly, while the engine was completely overhauled and tested by Laurentian Air Service Ltd.

in Ottawa. The restored aircraft was first flown by John Paterson at Fort William on a cold December 17, 1969. It was later flown to Ottawa with a commemorative load of mail over September 23-25, 1970, by Francis Kearney with stops at Duluth, Marquette (Michigan), Pellston (Michigan), Sarnia, Kitchener, Oshawa and Kingston. It arrived at Rockcliffe at 15:15 on the 25th for the hand-over ceremony.

In 1971 a Taylor E-2 Cub was purchased in the United States. While few E-2s were used in Canada, the type is of interest as it was the first Cub and the direct ancestor of the numerous and well-known Taylor/Piper J-2 and J-3 Cubs. In June 1985 the specimen was exchanged with Garth Elliot of Toronto for his meticulously restored E-2 Cub.

During a talk to the Toronto Chapter of the Canadian Aviation Historical Society, Father John MacGillvray, an RCAF padre, mentioned that he would sell his de Havilland D.H.80A Puss Moth on his approaching retirement from the RCAF. After the meeting, I suggested he offer it to

The Stearman 4EM purchased in its duster form and restored by John Paterson as a 4EM mailplane of Canadian Airways leaves Oshawa, Ontario, on September 25, 1970, for Rockcliffe where it will be presented to the Museum on that same day.
(KMM)

71

The Stits SA-3A Playboy was initially registered as CF-IGK and finished as shown in 1956 by its builder, Keith Hopkinson.

(J.F. McNULTY)

the Museum and he agreed. George Neal flew the machine to Ottawa from Summerside, Prince Edward Island, on May 1, 1976, and on May 2, 1976, Father MacGillvray flew his Puss Moth for the last time from Ottawa International Airport to Rockcliffe for the official hand-over ceremony.

Since the mid-1950s, a rapidly growing fleet of home-built aircraft has appeared on the Canadian aviation scene, but there was no representation in the collection until 1978 of this popular activity of sportsmen pilots. In that year Dr. Kernohan offered his Stits SA-3A Playboy, C-FRAD, originally registered CF-IGK-X, to the Museum. The machine had been completed by its builder, the late K.S. Hopkinson, at Goderich, Ontario, in 1956 and was the first postwar home-built aircraft to receive a flight permit. The vigour with which Hopkinson pursued the cause of the home-builder, already well established in the United States, did much to facilitate and popularize the growth of this activity in Canada. The Stits thus seemed an ideal representative of this popular class and it was purchased in July 1978.

In March 1965, the CWM bought a former RCAF Auster A.O.P.6, which was on the civil aircraft register, from Garant Aviation of Montmagny, Quebec. At No. 6 Repair Depot, Trenton, it was refinished as a British Auster A.O.P.6 in which a Canadian, Captain P.J.A. Tees, had won a DFC in Korea.

The Auster A.O.P. 6 on the right in this photo taken at Rivers, Manitoba, circa 1950, is the machine now held by the Museum. Restoration in these markings is planned.
(*FLIGHT* VIA L. MILBERRY)

The Helicopters

All the helicopters in the collection except one have been transferred to the Museum by the Canadian Armed Forces as they were retired from service. The exception is the Sikorsky R-4, the first production helicopter, which was obtained in exchange for a Japanese Aichi "Val" with the Planes of Fame Museum of Chino, California. The "Val" had been acquired in exchange for RCAF services to its owner, R. Diemert of Carman, Manitoba, for bringing the "Val" and some Zeros to Canada from the South Pacific. However, the "Val" had been put into flying condition with many non-original parts and techniques and did not

In 1974 the Museum was presented this Aichi D3A2 Type 99 "Val," a Japanese carrier dive bomber. It had been restored privately in Canada and flown with an American engine. As it did not fit the collection and the workmanship was not up to Museum standards, it was traded.

(KMM)

seem to fit the Museum's collection, so the exchange was made in November 1983.

One other rotating-wing aircraft is the Pitcairn-Cierva PCA-2 Autogiro. An autogiro may be described as a rotating wing aircraft with an unpowered rotor. Its significance in the development of vertical flight is considerable, as once the technology had been developed to make the autogiro successful, the practical helicopter was inevitable. The Museum at first dealt with the American owner, but a satisfactory arrangement to bring it to Canada could not be worked out, and it was given to the Connecticut Aviation Historical Association. Later it came to the Museum on loan and then was purchased on August 15, 1969.

More than a quarter of a century has elapsed since the formation of the National Aviation Museum but it is unlikely that any of those associated with it in its early days could have envisaged the growth that would occur and how many outstanding gaps in the collection would be filled. Nevertheless some still persist and much work remains to be done.

This is the earliest known photo of Rockcliffe Aerodrome, taken June 2, 1921,
looking southwest. A small dock had been installed behind the hangar,
and a Curtiss HS-2L flying boat, likely G-CYAG, is tied up there.
The small triangle in the centre of the runways was a code marking indicating
a commercial airharbour including a designation of runway length.

(PAC PA 62456)

The first hangar at Rockcliffe,
photographed in 1921.
This is one of the canvas-covered
Bessoneau hangars donated by
Britain to Canada in 1920.
A Bristol F.2B fighter, G-CYDP,
is in the doorway, one of two of the
type operated by the Air Board.
(DND RE-18027)

An Air Board Curtiss HS-2L,
G-CYEJ, being hauled up a
primitive slipway at Rockcliffe
in the fall of 1922. This would
have been the first slipway and
probably had just been installed.
(KMM/G. R. HUTT)

An Air Board D.H. 4, G-CYDL,
at Rockcliffe, July 21, 1923, in
an apparently staged photo of
an airmail flight to Toronto.
The canvas hangars did not
stand up well in the Canadian
climate and by 1923 the canvas
covering of the hangar had been
replaced by wood; the original
frame remained with its
characteristic shape.
(PAC PA 062293)

*In late August 1929 the RCAF Siskin aerobatic team was based at Rockcliffe
while appearing at the Central Canada Exhibition held annually in Ottawa.*
(PAC PA 062716)

*On the afternoon of March 12, 1930, W/C W.G. Barker, VC, then Sales Manager,
Fairchild Aircraft Ltd., was demonstrating a Fairchild KR-21, CF-AKR,
to the RCAF at Rockcliffe when he apparently lost control and crashed
on the ice of the Ottawa River. He was killed instantly and Canada lost
one of its leading WWI fighter pilots.*
(DND RE-74-166)

Rockcliffe, 1930. The new slipway and hangar built in 1928 are prominent while the old hangar is dwarfed beside it. Aircraft at the dock are a Canadian Vickers Vancouver II, a Fairchild FC-2 and a D.H. 60 Moth. Parked halfway up the slipway are a Bellanca CH-300 Pacemaker and a Fairchild FC-2; behind the new hangar is a Fairchild FC-2W. On the airfield is a civil Fairchild 71; and two Armstrong Whitworth Atlases are returning from a flight. Behind the trees on the right are a Canadian Vickers Vancouver I and a Vedette.

(DND RE 68-6343)

BELOW
Col. and Mrs. C.A. Lindbergh prepare to cast off in their Lockheed Sirius on July 31, 1931, to continue their flight north to the Orient. Mrs. Lindbergh is on the wing and Charles Lindbergh's head is barely visible in the rear cockpit.

(PAC PA 62884)

HC-6935
THE AIRCRAFT LINE.
FURIES IN FOREGROUND

At an air display on July 14, 1934,
the RCAF's Siskin aerobatic team
was a major attraction but
the stars of the display were
the Hawker Furies (foreground)
of the RAF's No. 1 Squadron.
(PAC PA 03142)

An overview of Rockcliffe
during the air display on July 14.
(PAC PA 134310)

All newly designed Canadian aircraft were tested for airworthiness at Rockcliffe until 1938. Here the prototype Noorduyn Norseman I is having its final tests on skis in January 1936; it had been tested on floats the previous November just before freeze-up. The newly built photographic building can be seen in the background.

(DND HC7457)

RCAF No. 110 (Army Co-operation) Squadron worked up at Rockcliffe to go overseas in late 1939 and early 1940. Its Canadian-built Westland Lysander IIs are seen there on January 30, 1940. This marked the beginning of operations from the south side of the airfield.

(PAC PA 63638)

An RCAF Boeing B-17F Flying Fortress IIA of No. 168 (Transport) Squadron takes off from Rockcliffe on December 15, 1943, to inaugurate mail delivery to the Canadian forces overseas. This view looking northwest shows the two earlier hangars, a control tower having been added to the 1928 hangar; a new, much larger hanger is just west of them.

(DND PL 23355)

The RCAF displays its first jet fighter, a Gloster Meteor F. III, on September 20, 1945.

(DND PL 37104)

*This photo, taken about 1960, shows the three WWII hangars
that housed the aircraft collection from 1964 to 1987.*

(DND PL 100089)

The Centennial Air Display, on June 6, 1967.
Museum specimens and RCAF machines flown in for the occasion are on display on
the tarmac. In the background are four historic machines about to take off for
a flying display; left to right, they are: Aeronca C-2, Sopwith 2F.1 Camel,
Nieuport 28 of C.H. Palen, Sopwith Snipe, and Fleet 16B Finch II.

(KMM)

R.W. BRADFORD

The ink was hardly dry on the agreement that allowed the use of the three World War II RCAF hangars at Rockcliffe Airport to house Canada's aeronautical treasures when it was realized that the fire-prone temporary buildings could be used for only a limited time. However, there was much excitement in 1964 when the National Aeronautical Collection opened its doors to the public and little opportunity for serious thoughts of a future building.

The overall guidance of the National Aeronautical Collection was provided through the National Aeronautical Policy Advisory Committee (known as NACPAC). The Committee represented various bodies with a vested interest in the future of the collection, including the National Research Council, National Aviation Museum, Department of National Defence, Canadian War Museum, Air Industries Association of Canada, and Ministry of Transport. Soon the question of new housing was to become prominent on its agenda.

In 1966, when the National Aviation Museum at Uplands was absorbed into the newly-formed National Museum of Science and Technology, supposedly for a period of ten years, NMST's aeronautical interests, including Rockcliffe, were managed by the nucleus of the old NAM staff under the title "Aviation and Space Division." In 1967, Dr. David Baird, founding director of NMST, successfully negotiated for control and management of the entire National Aeronautical Collection. NACPAC continued its advisory role for a number of years.

In the early 1970s, Dr. Baird encouraged the RCAF Memorial Fund's president, Air Marshal C.R. Dunlap, to join the Museum's effort to gain support for new housing for the NAC. The RCAF Memorial Fund originally planned to build a museum at the RCAF base at Trenton but had insufficient funds to proceed. A/M Dunlap was instrumental in gaining the attention of higher levels of government and assisted Dr. Baird in starting the process that finally resulted in serious attention to the preservation of the National Aeronautical Collection.

As early as 1971, Dr. Baird began to prepare proposals for new facilities based on curatorial input from the NAM staff and the recommendations of NACPAC. In January 1972, a proposal was submitted to the Board of Trustees of the National Museums of Canada, which agreed in principle and suggested NACPAC set up a working committee. A/M Dunlap, representing DND, and George Skinner of the Ministry of Transport were assigned to investigate sites for the proposed museum. For a while it appeared that the National Aviation Museum, as it was now unofficially called, might be located at Uplands Airport.

There was now a rising tide of interest in the preservation of the National Aeronautical Collection, and it had become obvious that there would be strong support from the aviation community. But if the aviation community was to be successful in its approach to government, it had to speak with one voice. That voice appeared from an unexpected source. In 1972, I was approached by Pat Boyce of Boyce Aviation Limited, with a plan to create what became known as the National Air Museum Society under the presidency of a Brampton lawyer, Blaine Bowyer. Mrs. Boyce succeeded in generating a high level of interest and the members of the Air Museum Society immediately rallied with a common goal. Such organizations as the Canadian Aviation Historical Society, the Royal Canadian Flying Clubs Association, the Canadian Owners and Pilots Association, the World War I Flyers and numerous others added their influence by joining.

Coincidental with this rising support was the concern expressed by military organizations such as the Royal Canadian Air Force Association, as well as a number of senior retired RCAF officers, that fire might consume the National Aeronautical Collection in the tinder-dry World War II hangars at Rockcliffe. Several disasters in the late 1970s, including the destruction of a number of aircraft in similar wooden hangars at Calgary, Alberta, and Carp, Ontario, served to add weight to the argument that if this outstanding collection was to be preserved, then something must be done as quickly as possible. Even greater emphasis was added to this concern with the destruction by fire in 1978 of the entire collection of the San Diego Aerospace Museum, which was housed in wooden structures.

Although Dr. Baird, A/M Dunlap and the National Air Museum Society were able to gain ministerial audience, the Government did not feel it was in a position to provide new housing, and it was suggested that, if the risk was intolerable, the collection be dispersed to various locations in Canada. This suggestion was immediately discarded as unacceptable. However, as the 1980s dawned, various positive factors came into play that were to have a profound effect on the future of the National Aviation Museum. Before he left NMST on December 31, 1981, Dr. Baird's final proposal to find a suitable home for NAM was received by an enthusiastic Board of Trustees under Dr. Sean Murphy, chairman, and Judge René Marin, vice-chairman.

Coming of Age

Early in 1982, the National Museums of Canada were delighted at the announcement by Prime Minister Pierre Elliott Trudeau that new facilities would be built for the National Gallery and the National Museum of Man (now known as the Canadian Museum of Civilization). but there was no mention of the National Aeronautical Collection. Fortunately, the Hon. Francis Fox, then Minister of Communications, to whom the National Museums of Canada reported, had a driving interest in the preservation of our national heritage in all its forms and was successful in convincing Cabinet that $20 million should be approved for the first of three phases of a new National Aviation Museum. His success brought overwhelming relief to the aviation museum and communities.

In 1982, I became Acting Director of the National Museum of Science and Technology for two years until a new director could be found. The focus for the site of the National Aeronautical Collection's new building had now shifted from Uplands to Rockcliffe. Uplands had become less attractive with the Ministry of Transport airport development plan, which moved the original suggested museum site to an inferior location.

The Rockcliffe site had several distinct advantages. First, it was a historic airport established in 1920 by the Canadian Air Board, indeed one of the first official airports in Canada. Rockcliffe's rich history included the vital aerial photography operations of the 1920s and 1930s, as well as its highly important role during World War II. Rockcliffe could handle both land and water-borne aircraft, the very feature that had made it an ideal location in Canada's early bush-flying operations. The site also appealed because of its natural beauty and because of plans by the National Capital Commission to link it with the Rockcliffe Parkway leading to downtown Ottawa and with the north-south Eastern Parkway (renamed Aviation Parkway in 1988). In addition, the National Aeronautical Collection's live museum policy between 1970 and 1982, had brought a successful flying program of Museum aircraft to nearly all provinces for a period of twelve years.

A key factor in maintaining the live airfield environment at the Museum site is the Rockcliffe Flying Club. In keeping with the National Capital Commission's plans to develop the waterfront area in front of the Museum, the Rockcliffe Flying Club plans to move to the south side of the active runway and to the east of the Phase I Museum facility. This partnership in preserving aviation activities on the site is vital to

Construction began in the spring of 1984. This view of the site highlights the position of the Museum within the triangular-shaped, existing runway system. The temporary buildings can be seen in the upper left; Ottawa is in the background. Note the slipway to the Ottawa River for water-borne aircraft.

both parties and, it is hoped, will become a permanent part of the Rockcliffe experience.

By 1985, the Department of National Defence had decided to vacate the airport site. In view of the plans of the National Museums of Canada, this could not have been more timely. As early as 1981, the architectural needs of the National Aeronautical Collection had been outlined to Michael Lundholm, Director of Architectural Services for the National Museums of Canada, who took a particular interest in the project.

With the approval of monies for the new Museum, it was Michael Lundholm's task to put the Museum's needs in a form that could be used by the Public Works Canada design team under Assistant Deputy Minister Guy Desbarats, design architect for the project, to bring into being this important first-phase facility. Both the National Museums of Canada and Public Works Canada architectural staff were fully aware of the restrictions imposed by the relatively low funding for the project, and this sensitivity was partly responsible for the imaginative delta-shape structure envisioned by Public Works Canada. Placing the long hypotenuse wall of the triangular shape along the northeast-southwest runway would save money on the tarmac surface for moving aircraft to and from the Museum. In addition, the configuration would recall the triangular shape of the British Commonwealth Air Training Plan airfields built across Canada during World War II.

The trusses for the tetrahedral structure were assembled on site, and then hoisted into place.

By positioning the new building alongside the STOL-Port buildings 193 and 194, the Museum's needs could be met on an interim basis until Phases II and III were built. The STOL-Port buildings remained from the experimental 1975 air service between Ottawa and Montreal. Building 193, with a total area of 1907 square metres, would house the conservation-restoration facility, as it had been the STOL-Port hangar and provided enough space for aircraft to be dismantled or restored. Building 194, with a total of 464 square metres, would house the administrative, curatorial, research, archival, and library facilities for the National Aviation Museum.

Public Works Canada had originally thought that the Phase I building could be built to provide a total of 16 027 square metres of space, but budgetary considerations reduced this to 13 138 square metres by the time the sod-turning ceremony took place on May 25, 1983. The ceremony attracted much attention and, appropriately, the event was saluted by one of the vintage aircraft from the National Aviation Museum flown by the Museum's Chief Pilot (since 1979), George Neal.

The Shaping of a Dream

The design problem of creating housing for over 100 aircraft and thousands of artifacts could be best described as monumental. A

number of large and weighty aircraft in the collection dictated clear spans of 45 metres, including the hangar doors, and a height of 12 metres. These demanding requirements were met by a tetrahedral space-frame structure. Thomas Fuller Construction won the construction contract and their engineering consultants, Adjelian Rubeli Limited, used computer analysis and destructive testing of joints to come up with the ingenious design. The enormous roof structure is supported by six interior columns of pyramidal configuration with 26 gravity columns and 31 wind columns along the three perimeter walls. The narrowest point of the inverted pyramidal space-frame structure supporting the roof sits on an upright pyramidal concrete structure with footings resting on a number of large steel pipe columns driven into the ground, in some cases, as deep as 52 metres. The narrowest point of the intersection of the pyramids is at a height off the floor comparable with the large aircraft wing spans, thereby allowing the 45-metre clearance.

The roof structure was assembled from 240 steel trusses to make up the 26 sections. The general contractor awarded the supply-fabricate-erect subcontract to F.L. Metal Inc. of Montreal. They, in turn, engaged the skills of Dulepka Equipment Limited of Ottawa for on-site assembly and erection of all the trusses assembly including on-site welding. The president of the company, Robert Dulepka, personally oversaw the work and orchestrated the delicate lifting of the 26 sections ranging in

OPPOSITE
The "spine" can be seen in the foreground in this spectular view of the Museum as it takes shape.
(J. DORN NAM 17382)

ABOVE
Aircraft rest on the tarmac outside the new building, adding the final touch to its attractiveness from the air.
(J. DORN)

The tetrahedral structure is not only practical, but evokes a feeling of space and beauty. Note the inverted and upright pyramidal structures upon which the roof rests.

Work crews are dwarfed by the WWII Lancaster bomber as it is rolled into its new home.

weight from 9000 to 72 500 kg. As many as eight cranes were used on the larger lifts, four attached to the actual assembly, two for lifting, turning over and positioning the inverted pyramid columns once the assembly was up in the air, and two other cranes for handling fill-in steel for the assembly.

The project was so unusual that it twice caught the eye of CBC Television, which recorded the delicate operations as Dulepka used two-way radios to communicate with the crews of the cranes. The resulting structure is probably the largest welded tetrahydral space-frame structure in Canada, a fact which would gladden the heart of Dr. Alexander Graham Bell, the great inventor and aeronautical experimenter, who in 1904 conceived and patented in the United States the tetrahedral system for cellular kite structures. Three years later, he built a tetrahedral tower on his estate near Baddeck, Nova Scotia to prove its use in civil engineering. It is fitting, therefore, that the first aircraft in the "Walkway of Time" within the Museum building is the replica of the *Silver Dart* that flew February 23, 1959, at Baddeck to commemorate the flight of the original *Silver Dart* in 1909, a historic event brought about by the encouragement of the Aerial Experiment Association founders, Dr. Bell and his wife, Mabel.

Stretching its wings, The Falcon *rests on the grounds of McGill University in Montreal, its original home.*
(J. DORN NAM 18655)

The west wall of the museum building houses what is referred to as the spine, the exterior of which is covered with white porcelain enamel panels. The rest of the building is faced in silvery-grey aluminum panels. The white porcelain enamel steel panels are also used to effect in the Museum's semi-circular foyer and the introductory exhibits area. The foyer has, as its focal point, a full-scale bronze reproduction of Dr. Tait McKenzie's classic 1932 work *The Falcon*. This sculpture serves as a symbol of the ancient human desire to fly.

The spine also houses the gift shop, cloak room, washrooms, first aid station, security offices, information desk, various electrical and systems control rooms and a temperature controlled area for sensitive artifacts. On the inside wall of the spine, there is a mezzanine with viewing galleries that allows visitors to overlook the building's displays. Immediately past the introductory display area, a large cylinder with an 80-square-metre floor area rises to the roof structure, which houses the impressive Royal Canadian Air Force "Hall of Tribute," a bronze and stainless steel work of art created by Toronto sculptor John McEwen. The external surfaces are clad in porcelain white steel panels consistent with the main entrance and the exterior of the western wall. The Hall of Tribute is intended to be a quiet place of remembrance dedicated to the men and women who served in the Royal Canadian Air Force.

Since this unusual building is on an active airfield, we wanted the grey and white roof to be attractive from the air, a requirement that dictated that all systems, such as sprinkler lines and electrical conduits,

From the mezzanine level the visitor can view the aeronautical "attic." One of the theatres is taking shape just beyond the stairs.

and heating and air conditioning ducts, be contained within the space frame structure. The roof design underwent wind tunnel tests to ensure that the prevailing winds adequately dispersed the snow. Although the lighting system is primarily artificial, natural daylight is admitted through skylights and the glazed upper areas of the north and west walls. Ultraviolet screens protect the collection from the harmful effects of ultraviolet rays.

The Phase I building was designed for display purposes but will have to be used for both display and storage until Phase II comes into being. Despite this constraint, a little over 6800 square metres of displays, highlighting 43 aircraft along with numerous artifacts, power-plants and interpretive displays, have been put in place in the chronologically ordered Walkway of Time. From the ground level or the mezzanine, visitors may view the aeronautical "attic."

The interior structure of the Museum is finished predominantly in off-white with some light grey areas. Coursing through the arrange-

ment of aircraft is a medium grey carpet to guide the visitor. A 40-seat theatre highlighting Canada's bush-flying era is situated near the spine between the display and storage areas.

The future Phase II building will have an enlarged conservation-restoration workshop, a large storage area that will remove the stored aircraft from Phase I and free it for expanded displays, a restaurant, curatorial offices, an auditorium and a glazed walkway with a view into the restoration and storage areas. Phase III will be a smaller, mirror image of Phase I and will house additional displays of a more current or specialized nature. The total area of the three phases will be about 24 000 square metres.

The National Aviation Museum at Rockcliffe, in its present and future stages, will provide Canadians and visitors from other lands an experience that will emphasize the importance of aviation to this country in peace and war, a heritage we can be proud to pass on to future generations.

The National Aviation Museum proudly awaits its first visitors.
(MALAK)

PART TWO

The Collections

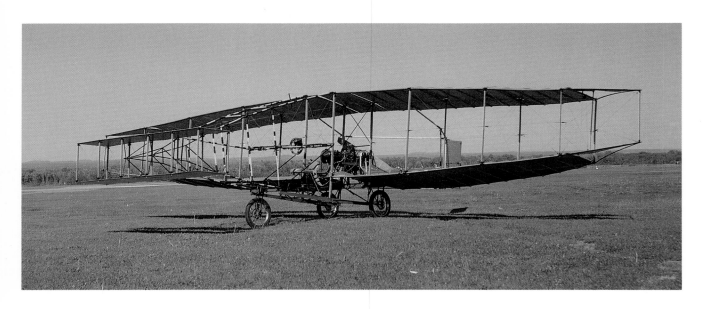

A.E.A. **Silver Dart**

SPECIFICATIONS

ORIGINAL AIRCRAFT

Powerplant: One 40 hp Curtiss

Wing span,
upper and lower: 49 ft 1 in (15 m)

Length: 39 ft 4 in (12 m)

Height: 9 ft 7 in (2.9 m) approx.

Wing area: 420 sq ft (39 m²)

Weight, empty: 610 lb (277 kg)

Weight, gross: 860 lb (390 kg)

Performance: not available.

The *Silver Dart* was the fourth and most successful powered aircraft designed and built by the Aerial Experiment Association. The A.E.A. was started by Dr. Alexander Graham Bell and backed by Mrs. Bell, and a member was appointed to serve as project engineer for each machine. J.A.D. McCurdy was responsible for the *Silver Dart*.

The aircraft followed the general design practices established on the preceding machines but a 40 hp water-cooled Curtiss engine was installed. This was Curtiss' second water-cooled engine, and a marked improvement over the air-cooled engine in the previous aircraft. Rubberized fabric was used to cover the aircraft's surfaces and it was the silvery appearance of one side of this fabric that caused McCurdy to give the aircraft its name.

The *Silver Dart* was built at Hammondsport, New York, and first flew there on December 6, 1908, with McCurdy as pilot. Further flights there followed; then Dr. Bell had it shipped to Baddeck, Nova Scotia, on January 6. The *Silver Dart*, piloted by McCurdy, made the first powered flight in Canada on February 23, 1909, and went on to make the first passenger flight in Canada at Petawawa, Ontario, on August 2 before being written off in a crash the same day.

The Museum holds two *Silver Dart* reproductions. LAC Lionel

McCaffery, RCAF, was the instigator of the scheme to build and fly a reproduction to mark the 50th anniversary of its historic first flight. The idea was taken up by the RCAF and two reproductions were made at No.6 Repair Depot at Trenton, one with a wooden duplicate of the original Curtiss engine and the other with a modern 65 hp Continental engine. W/C Paul A. Hartman first flew the reproduction at Trenton on January 29, 1959. After several flights there the aircraft was shipped to Baddeck on February 5. It was first flown from the ice on February 21, and this was the best of the ten flights it made. A ceremony was planned for the 23rd, the anniversary date, with various dignitaries in attendance. A gusty wind arose that day but nevertheless a flight was attempted, which unfortunately ended in a crash, damaging the machine. The pilot was not hurt.

The aircraft was repaired and served as the sole full-scale machine on exhibit upon the opening of the National Aviation Museum on October 25, 1960. The second aircraft was held briefly by the RCAF before being transferred to the Museum.

A.E.G. G.IV

SPECIFICATIONS

Powerplant: Two 260 hp Daimler Mercedes D.IVa

Wing span, upper: 18.30 m (60 ft)

Wing span, lower: 17.30 m (56 ft 9 in)

Length: 9.7 m (31 ft 10 in)

Height: 3.78 m (12 ft 5 in)

Wing area: 74 m² (796 sq ft)

Weight, empty: 2488 kg (5,486 lb)

Weight, gross: 3664 kg (8,079 lb)

Speed, max.: 165 km/h (103 mph)

Initial climb: 1000 m (3,280 ft) in 5 min

Ceiling: 4500 m (14,760 ft)

In late 1916 the Allgemeine Elektrizitäts Gesellschaft (General Electric Co.) of Hennigsdorf, Berlin, Germany, first flew its new bomber, the G.IV. The type went into general use with its contemporaries, the Friedrichshafen G.III, the Gotha G.IV and G.V, as a standard twin-engine bomber during 1917 and 1918. It was used on the western front, in Austria against the Italians, in Macedonia and in Salonica.

Its design followed the pattern established by its predecessors, the

A.E.G. G.I, G.II and G.III. It normally carried a crew of three, armed with two Parabellum machine guns and a bomb load of 780 kg (1,720 lb) with bombs ranging up to 300 kg (662 lb) in size. As its range was shorter than that of its contemporaries, it tended to be used for tactical bombing behind the lines by both day and night rather than for long-range strategic bombing.

The large white numeral IV on the fuselage and the lack of evidence of use suggest that the Museum's specimen had just been assigned to a German unit, possibly a training unit, before it was captured. Shipped from Dieppe, France, on the SS *Venusia* on May 23, 1919, it was one of many war trophies brought to Canada that year (see Appendix 1). It was taken to Toronto but there is no evidence that it was displayed there, as some other German aircraft were.

Its movements over the next 40 years are not well recorded but it went to Camp Borden with the other war trophy machines and managed to escape destruction there. It was subsequently moved to Ottawa and in 1939, along with the Junkers J.1, was offered to the Aeronautical Museum at the National Research Council; the two machines were stated to be "now in the Archives." The offer was turned down for lack of space. The aircraft is known to have been stored in a warehouse operated by the Canadian War Museum in the 1950s, apparently having come to the CWM through "right of possession" from the War Trophies Board.

In 1968 it was sent by the CWM to No.6 Repair Depot, Trenton, for restoration by the RCAF, which was completed in 1969. The engines and nacelles were not with the aircraft and two 160 hp Daimler Mercedes D.III engines were installed until the rare Mercedes D.IVa engines could be obtained, but the radiators and cowlings were not built and installed. A Mercedes D.IVa, possibly a unique specimen, has since been obtained and it is hoped that reconstruction of the nacelles will shortly go ahead.

The A.E.G. G.IV is a unique specimen as it is: A/ the only one of its type extant; B/ the only twin-engine German World War I aircraft extant; and C/ the only surviving example displaying the unique German World War I night lozenge fabric.

Aeronca C-2

The Aeronca C-2 had its origin in the single-seat Roché light plane design of Jean A. Roché, an engineer with the US Army Air Service at McCook Field. It was built by Roché and John Dohse and was first flown by John Dohse on September 1, 1925, at Dayton, Ohio.

Roché wanted to create a light aircraft that was economical and

SPECIFICATIONS

Powerplant: One 36 hp
        Aeronca E-113
Wing span: 36 ft (11 m)
Length: 20 ft (6 m)
Height: 7 ft 6 in (2.3 m)
Wing area: 142.2 sq ft (13.2 m²)
Weight, empty: 406 lb (184 kg)
Weight, gross: 700 lb (317 kg)
Speed, max.: 80 mph (129 km/h)
Cruising speed: 65 mph (105 km/h)
Service ceiling: 16,500 ft (5030 m)

practical, and he followed glider practice in doing so. While the Roché light plane achieved his objective, it was not immediately put into production. When the Aeronautical Corporation of America was formed late in 1928, it bought the Roché aircraft and retained its designer as a consultant. The company modified the design to facilitate production and introduced its own engine, made for it by the Govro-Nelson Co. of Detroit.

The Aeronca C-2, construction number 2, was first flown on October 20, 1929, and was introduced to the public at the St. Louis Air Show in February 1930. It sold originally for US$1,555 and was reduced in 1931 to $1,245. The C-2 and the very similar C-3 two-seater proved popular, even in those Depression years, and are generally credited with triggering the light plane boom in North America.

The first C-2 appeared in Canada at a Montreal air meet in September 1930. Dougal Cushing of Montreal bought it on the spot, believing it had a great future in the Canadian flying clubs, but this was not to be, as government assistance to the clubs was conditional on their aircraft being British or Canadian made. Nevertheless the Aeronautical Corporation of Canada was formed at Toronto and imported and sold 17 C-2s and C-3s during the 1930s.

The Museum's specimen, construction number 9 and the eighth C-2 made, was originally sold to G.A. Dickson of Pittsburgh and passed through the hands of several private owners. During this time its original vertical tail was replaced by that of a late production C-3 and its original Aeronca E-107 engine of 26 hp was replaced by the E-113 model of 36 hp. The Museum purchased it in February 1967 and immediately restored it to flying condition; it flew during the Air Force Day celebration at Rockcliffe on June 10 of that year.

Its Canadian registration of CF-AOR corresponds to that of M. Foss's C-2 in the 1930s and its finish is as it was in the mid-30s. Its wire-braced wings and light appearance continue to arouse the interest of Museum visitors.

Auster A.O.P. 6

SPECIFICATIONS

Powerplant: One 145 hp D.H. Gipsy Major 7
Wing span: 36 ft (11 m)
Length: 23 ft 9 in (7.2 m)
Height: 8 ft 4-1/2 in (2.6 m)
Wing area: 184 sq ft (17 m²)
Weight, empty: 1,469 lb (666 kg)
Weight, gross: 2,210 lb (1002 kg)
Speed, max.: 122 mph (196 km/h)
Initial climb: 660 ft/min (200 m/min)
Service ceiling: 12,000 ft (3650 m)

The Auster aircraft had their beginning with the formation of Taylorcraft Aeroplanes (England) Ltd. in 1938 to make the American Taylorcraft aircraft under licence for the private flying market. The Taylorcrafts were designed by C.G. Taylor, who had earlier been responsible for the design of the well known Taylor (later Piper) Cub. During World War II some Taylorcrafts were ordered by the British Army Air Corps for observation work. The design was developed and improved for military use, the D.H. Gipsy Major 1 was adopted as the powerplant, and the name Auster was given to the aircraft.

The company was re-formed in 1946 under the name Auster Aircraft Ltd. and continued to make light aircraft for civil and military use, including new designs of its own. In 1948 Canada ordered 34 Auster 6s, which were a development of the original Auster 1. They were used for army observation and were stationed at Rivers, Manitoba, until withdrawn from service in 1958. At that time they were sold for civil use in Canada and many continue in private hands today.

In 1965 the Canadian War Museum purchased Auster 6 CF-KBV, ex-RCAF 16652, and had No.6 Repair Depot of the RCAF recover the machine and finish it in the markings of an RAF Auster A.O.P. 6, VF582, which a Canadian, Capt. P.J.A. Tees, had flown in Korea, winning a DFC.

Avro 504K

SPECIFICATIONS

Powerplant: One 110 hp Le Rhône
  or one 130 hp Clerget
Wing span, upper and lower: 36 ft
  (11 m)
Length: 29 ft 5 in (9 m)
Height: 10 ft 5 in (3.2 m)
Wing area: 330 sq ft (31 m²)
Weight, empty: 1,231 lb (558 kg)
Weight, gross: 1,829 lb (829 kg)
Speed, max.: 95 mph (153 km/h)
Climb: to 3,500 ft (1070 m) in 5 min

The prototype of the famous 504 series of Avro aircraft first flew in September 1913. The type entered military use as a standard service machine for reconnaissance by the Royal Flying Corps and for anti-Zeppelin patrols by the Royal Naval Air Service. With its development as a trainer by Major R.R. Smith-Barry in the 504J and 504K versions, the type came into the role that earned it enduring respect and affection in many quarters. In the postwar period the 504 saw military and/or civil use in many countries around the world, which served to further enhance its reputation.

A Canadian version was just coming into production in Toronto when the war ended and only two were made. On the formation of the Canadian Air Force in 1920, the 504K was adopted as its standard trainer and the type remained in use with its successor, the RCAF, until 1928. The RCAF machines were part of the Imperial Gift of 1920. Only six 504s came into Canadian civil use but one of them made the first winter flight to James Bay in February 1922.

The Avro 504K, 130 hp Clerget powered, in typical RCAF markings of 1924-25, is on permanent display at Rockcliffe. (KMM/H. TATE)

The Avro 504K, 110 hp Le Rhône powered, in interim RCAF marking scheme used on three aircraft briefly in 1921-23, is on loan to the Western Canada Aviation Museum, Winnipeg.
(KMM/H. TATE)

Avro 504K, 110 hp Le Rhône powered, in typical Air Board markings of 1921 but with roundels added by the RCAF for their 1967 flying display.
(KMM/H. TATE)

Avro 616 Avian IVM

The three Museum specimens were acquired in the following way. In 1967 the RCAF was planning a series of air displays at major centres from coast to coast as part of the Centennial celebrations. The air force wanted to fly an example of an early RCAF aircraft at these displays and selected the Avro 504K, its first trainer. One aircraft, now (G-CY)FG in the Museum, was bought from C.H.C. Palen of Rhinebeck, New York, who had been using it for flying exhibitions; it had earlier been in the Roosevelt Field Museum, New York. The RCAF bought a second aircraft from Major J. Appleby in California, which had earlier been in Connecticut. This aircraft, now G-CYCK, was at the time partially restored.

The RCAF then decided to finish the restoration of Major Appleby's aircraft and build a completely new 504K, G-CYEI. These two machines, powered with 110 hp Le Rhône engines, did all the RCAF exhibition flying. The Palen 504K was repaired where necessary, recovered and fitted with a 130 hp Clerget engine, as were all early RCAF examples, and retained for Museum exhibit.

After appearing in the 1967 RCAF flying displays and treating many Canadians to the nostalgic sight of a 504K airborne and the smell of burnt castor oil, the machines were retired to the RCAF historic aircraft collection, which became part of the National Aviation Museum. G-CYCK is used in flying displays.

*

The original Avian, the Avro 581, appeared as a prototype in 1926 and, slightly modified, was flown to Australia from England in 1928 by H.J.L. Hinkler. A new production version appeared in 1927 designated the Avro 594 Avian. This was very similar to, and a direct competitor of, the de Havilland D.H.60 Moth, and like the Moth it was powered with several different types of engines.

Avro followed de Havilland's lead in 1929 by bringing out a metal-fuselage Avian, which they designated the Avro 616 Avian IVM, and it too was powered with several different engines. Both the Avro 594 and 616 Avians were used widely by private owners, civil operators and flying clubs, and several long-distance flights were made in the types. No military use was made of the 594 other than experimental, and the South African Air Force and the RCAF were the only military organizations to use the 616. So while the Avians were built and used in good numbers, they never matched the quantity of their competitor, the D.H. Moth. The 616M was built under licence in the United States

SPECIFICATIONS

Powerplant: One 100 hp Armstrong-
Siddeley Genet Major
Wing span, upper and lower: 28 ft
(8.5 m)
Length: 24 ft 3 in (7.4 m)
Height: 8 ft 6 in (2.6 m)
Wing area: 245 sq ft (23 m²)
Weight, empty: 1,000 lb (454 kg)
Weight, gross: 1,600 lb (726 kg)
Speed, max.: 100 mph (160 km/h)
Ceiling: 12,500 ft (3810 m)

by the Whittlesey Manufacturing Co. in Connecticut and some components were made in Canada by the Ottawa Car Manufacturing Co.

Twelve wooden-fuselage 594 Avians were imported for civil use. One of them entered Canadian aviation history books after being flown on wheels in January 1929 from Edmonton to Fort Vermilion, Alberta, by W.R. (Wop) May to deliver diphtheria anti-toxin. Thirty-nine Avro 616 Avians were delivered by the Ottawa Car Manufacturing Co. and some of these British machines had wings and tail surfaces made by the Canadian company. Two were delivered to private owners and the balance to the RCAF or the Department of National Defence for use by the flying clubs.

The Museum's specimen was delivered to the RCAF as RCAF 134 in April 1930 and was passed on by the DND as CF-CDQ in 1932 to the Moose Jaw Flying Club, and then on to the Winnipeg Flying Club in 1935. Sold to a private owner in 1937, it was placed in storage during the war and afterwards sold to Charles Graffo of Winnipeg, whose estate donated it to the Museum in 1968. The specimen is complete but needs restoration to bring it to good display condition.

Avro 652A Anson V

SPECIFICATIONS

Powerplant: Two Pratt & Whitney
        Wasp Jr. R985-AN-14B
Wing span: 56 ft 6 in (17.22 m)
Length: 42 ft 3 in (12.9 m)
Height: 13 ft 1 in (4 m)
Wing area: 410 sq ft (38 m²)
Weight, empty: 6,693 lb (3035 kg)
Weight, gross: 9,275 lb (4207 kg)
Speed, max.: 190 mph (306 km/h)
Climb: 8,000 ft (2440 m)
      in 9 min 15 sec
Service ceiling: 20,550 ft (6260 m)

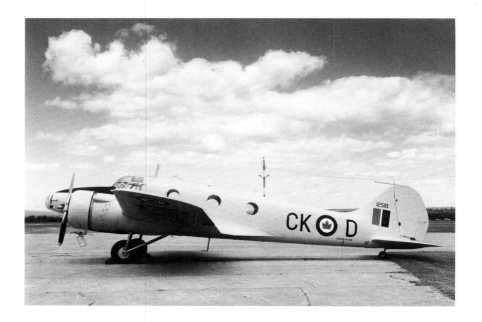

The original Anson was designed and built as a civil aircraft for Imperial Airways Ltd. as the Avro 652 and was first flown in January 1935. Its design was based on that of the Fokker F.VII/3m but converted to a low-wing monoplane. Only three 652s were made but the Royal Air Force issued a requirement for a military adaptation of the type and it entered quantity production as the 652A. Many were used for coastal defence patrol until the early part of World War II, and Ansons were credited with destroying six enemy aircraft before being retired from these duties.

When the British Commonwealth Air Training Plan was set up in Canada in 1939, the Anson was selected as the standard twin-engine trainer. The aircraft were to be built in England, except for the wooden wings, which were to be made in Canada. Events in Europe in 1940 made it necessary to build the Anson in Canada. A new version, the Anson II, was created for Canadian production with 330 hp Jacobs engines and a powered hydraulic undercarriage. A further Canadian development, the Anson V, was put into production in 1943 with Pratt & Whitney Wasp Jr. engines and a moulded plywood fuselage to replace the fabric-covered one of the earlier models. This enhanced the performance of the aircraft and greatly increased the comfort of the crews and the student pilots and navigators. The lion's share of twin-engine pilot and navigator training in the BCATP was carried out on Anson IIs and Vs.

After the war, the Anson Vs were soon phased out of the RCAF but many entered civil use and some were exported to at least eight countries. They were used as light transports, and their all-wood construction made them popular for magnetic surveying.

Anson Vs were made in Canada by both Canadian Car and Foundry Ltd. at Amherst, Nova Scotia, and by MacDonald Bros. Aircraft Ltd. at Winnipeg, Manitoba. The Museum's specimen was one of 748 built by MacDonald Bros. and was accepted by the RCAF in June 1945. It served with Air Transport Command from 1948 to 1952 and at the Central Experimental and Proving Establishment from 1952 to 1954. It was then stored until selected for preserving as a Museum aircraft in 1964.

Avro 683 Lancaster X

SPECIFICATIONS

Powerplant: Four 1,620 hp
    Rolls-Royce
    (Packard) Merlin 224

Wing span: 102 ft (31.1 m)

Length: 69 ft 6 in (21.1 m)

Height: 20 ft 4 in (6.2 m)

Wing area: 1,300 sq ft (121 m²)

Weight, empty: 35,240 lb (15 985 kg)

Weight, gross: 65,000 lb (29 484 kg)

Speed, max.: 272 mph (438 km/h)

Initial climb: 910 ft/min (277 m/min)

Service ceiling: approx. 24,700 ft
    (7530 m)

The Avro Lancaster originated with the Avro 679 Manchester, which was powered by two Rolls-Royce Vulture engines and first flew in July 1939. Difficulties with the Vulture caused the Air Ministry to stop its production after only 200 Manchesters were built. A.V. Roe, on its own initiative, then designed a four-engine development of the Manchester which became the Lancaster, the most successful night bomber of World War II. The first Lancaster, in fact a modified

Manchester, was first flown in June 1941 and was then ordered into production.

Lancasters were first issued to RAF squadrons at the end of 1941 and were first used in operations in March 1942. An order for Canadian-built Lancaster Xs was given in November 1941 and the first Mk.X was flown on August 1, 1943, at Malton, Ontario. The outstanding feature of the Lancaster was its 33-foot, unobstructed bomb bay, which could carry bombs up to 8,000 lb (3630 kg) in weight; some specially modified aircraft carried special-purpose bombs of up to 22,000 lb (9070 kg). Canadian Mk.Xs were similar to the British Mk.IIIs, the main differences being that the Mk.X had a different electrical system and the last 272 Mk.Xs had a Martin mid-upper turret fitted with 0.50 guns, giving increased fire power. Canadian Lancasters were built by Victory Aircraft at Malton, Ontario.

All Mk.X Lancasters served with RCAF squadrons 419, 420, 425, and 428 of No.6 Group. They took part in bombing operations against Germany and France as well as minelaying operations. In the course of these operations Pilot Officer A.C. Mynarski won a posthumous Victoria Cross for bravery during an attack on Cambrai, France. In the postwar years Lancasters were modified to serve in various RCAF roles, including photo reconnaissance, maritime patrol, search and rescue, and navigator training. They were finally retired in 1965 after 20 years of peacetime service.

The Museum's specimen, KB944, is very close to its original wartime condition and is the only Mk.X extant in this condition. It was flown to England in March 1945 and was delivered to 425 Squadron in May, just too late to take part in bombing operations. It returned to Canada later in 1945 and spent most of the following years as a stored reserve aircraft, with only a brief period with 404 Squadron at Greenwood, Nova Scotia, before it was placed in the RCAF's historic aircraft collection in 1964.

The aircraft was refinished by the RCAF in 1964 in the markings of KB760, which was aircraft P of 428 Squadron. This was an earlier Mk.X fitted with a Frazer Nash mid-upper turret.

Avro Canada CF-100 Mk.5

SPECIFICATIONS

Powerplant: Two Avro Orenda 11
7,300 lb st
(53 200 kg st)
Wing span: 57 ft 2-3/5 in (17.4 m)
Length: 54 ft 1-3/4 in (16.5 m)
Height: 14 ft 6-2/5 in (4.4 m)
Wing area: 591 sq ft (54.9 m²)
Weight, empty: 23,100 lb (10 478 kg)
Weight, gross: 33,528 lb (15 208 kg)
Speed, max.: 481 knots at 30,000 ft
(890 km/h at 9144 m)
Climb: 8,750 ft/min (2670 m/min)
Combat ceiling: 45,000 ft (13 720 m)

The CF-100 was designed and built by Avro Canada at Malton, Ontario, to meet an RCAF requirement issued in 1946 for a two-seat all-weather fighter. The two prototypes, the Mk.1s, were powered with Rolls-Royce Avon engines in order to test the aircraft as soon as possible pending completion of Orenda engines being developed by the Engine Division of Avro Canada. The prototype was first flown on January 19, 1950, by W.A. Waterton at Malton. The type was officially named Canuck by the RCAF but the name was never felt appropriate and did not come into general use.

The type was developed through four more Mk. numbers, together with some sub-types which served different roles with different armament, and with different versions of the Orenda engine. The two most numerous versions were the Mk.4 and 5. The Mk.4 had Orenda 9 or 11 engines, APG-40 radar and MG-2 fire control system and was armed with an eight-gun fuselage pack and two 29-rocket tip pods. This was followed by the Mk.5 high-altitude version with Orenda 11 engines and extended wings and two 29-rocket tip pods.

Although not quite as fast as smaller contemporary fighters, its excellent climb, fine radar and fire control systems, twin-engine reliability and all-weather capability made the CF-100 probably the best all-weather fighter of its time and ideal for the defence of northern Canada. In 1957 Belgium selected the type over its competitors and obtained the CF-100 Mk.5 for its air force.

The RCAF CF-100s were stationed at North Bay, Ontario; Bagotville, Quebec; Comox, British Columbia; and Uplands (Ottawa). In 1956 four squadrons of CF-100s replaced four North American Sabre squadrons in No.1 Air Division, serving in France and West Germany, where their all-weather capability was much appreciated in the fog and murk of Europe. When retired from their fighter role, some CF-100s were fitted with electronic counter measures (ECM) equipment and others were fitted out as target tugs. All CF-100s had retired from RCAF service by September 1981.

Two CF-100 Mk.5s are held by the Museum, 100757 (ex-18757) and 100785 (ex-18785), and both are fitted with tip tanks. Before being presented to the Museum, 100785 was finished in black by the Canadian Armed Forces to resemble the CF-100 Mk.1s. Both aircraft had served with the electronic warfare unit, 414 Squadron.

Bell HTL-6 (47G)

SPECIFICATIONS

Powerplant: One 200 hp Franklin
    6V4-200-C32

Rotor diameter: 35 ft 1-1/2 in (10.71 m)

Length: 41 ft 2-1/2 in (12.56 m)

Height: 9 ft 6 in (2.9 m)

Weight, empty: 1,380 lb (626 kg)

Weight, gross: 2,350 lb (1066 kg)

Speed, max.: 90 mph (145 km/h)

Speed, cruising: 77 mph (124 km/h)

Inclined climb: 800 ft/min
    (244 m/min)

The Bell Model 37 helicopter was made originally by Bell Aircraft Corp. of Buffalo, New York, and later by the Bell Helicopter Corp. of Fort Worth, Texas. It resulted from much experimental work done on the Model 30, which was first flown in 1943. The first Model 47 was made later in 1945. In 1946, the type became the first helicopter to be commercially licensed and, in January 1947, the first commercial helicopter to be delivered. Since then the Bell 47 has been produced in many versions and its production continued until 1973 in the USA. The 47G and 47J were also made in Italy by Construzioni Giovanni Agusta during the years 1954-74. Westland Aircraft Ltd. produced the 47G in

England, where it was known as the Sioux, while in Japan it was made by Kawasaki Heavy Industries Ltd. Two versions were named by Bell, with the 47H being christened the Bellairus and the 47J the Ranger. Altogether about 6,439 of the various versions of the Bell 47 were made by Bell and its licensees.

The Model 47 was basically a three-seater, although some two- and four-seat versions were made, and it was classified as a light helicopter. It entered civil, military and naval service in many countries. Its duties comprised the whole range of activities carried out by light helicopters and included training, surveying, agricultural spraying, light transport, ambulance duties, rescue work, and police patrol. Many continue in active use today. It is one of the most successful helicopters built and is probably the most numerous of all helicopters.

The Bell 47 was introduced by the Royal Canadian Navy in August 1951 when three 47D-1s were taken on strength. The first helicopters to be operated by the navy, they were designated HTL-4s in the RCN, following the US Navy's system. The designation of the Museum specimen, HTL-6, was the USN designation for the Bell 47G and was also adopted by the RCN. The specimen was given RCN serial 1387 and taken on strength in June 1955, joining the three HTL-4s already in service. It served with helicopter squadron HU-21 based at HMCS *Shearwater* at Dartmouth, Nova Scotia. During its naval career it carried out many of the duties mentioned above, including service on patrol ships on survey work along the coasts and in the Arctic. It was retired to the Museum in 1966.

Bellanca CH-300 Pacemaker

The family of Bellanca monoplanes which began with the WB-1 prototype in 1925 was used in many long-distance flights in the 1920s and 1930s, sharing honours in this activity with the Lockheed Vega. The second prototype, the WB-2 *Columbia*, piloted by Clarence Chamberlin, flew from New York to Germany two weeks after Lindbergh's flight, and in fact it was only because he could not obtain the *Columbia* that Lindbergh commissioned the construction of the famed *Spirit of St. Louis*. The *Columbia* flew the Atlantic again in 1930 with Capt. J. Erroll Boyd at the controls, which made it the first aircraft to complete two Atlantic crossings, and Boyd became the first Canadian to make the crossing. Other notable long-distance attempts were made in Bellancas piloted by Williams and Yancey, Hillig and Hoiris, Boardman and Polando, Pangman and Herndon, and others.

What made the Bellanca family the choice for these flights was its

SPECIFICATIONS*

Powerplant: One 300 hp
        Wright J-6-9
Wing span: 46 ft 4 in (14.1 m)
Length: 27 ft 9 in (8.5 m)
Height: 8 ft 4 in (2.5 m)
Wing area: 273 sq ft (25.4 m²)
Weight, empty: 2,275 lb (1032 kg)
Weight, gross: 4,072 lb (1847 kg)
Speed, max.: 140 mph (225 km/h)
Climb: 1,100 ft/min (335 m/min)
Service ceiling: 18,000 ft (5490 m)

* Specifications are given with the
  original engine type installed and not
  with the Wasp Jr. since fitted.

remarkable lifting capacity; a Packard-diesel-powered Bellanca made an unrefuelled endurance flight in 1931 of 84 hours 33 minutes, the most enduring of aviation records, not surpassed until the Rutan Voyager's 216-hour world flight in 1986. The Bellancas' weight-lifting capability made them a natural choice for bush flying, and although less powerful than their Pratt & Whitney Wasp-powered contemporaries, the Bellancas could give them close competition when it came to lifting a load of freight. They were initially imported into Canada; later six were built by Canadian Vickers Ltd. at Montreal. The first Canadian Bellanca was sold to a private owner, A.S. Dawes of Montreal, in 1929, and bush operators and the RCAF bought others. The RCAF had 13 Bellancas, used mainly for photography. Canadian Airways operated eight, Commercial Airways owned five and General Airways owned four and leased one. Some of these machines were the same aircraft under different ownership.

The Museum's specimen was built in 1929 and sold to El Paso Air Service of El Paso, Texas. Then it served with two private owners in California from 1933 to 1936, when it was sold to Marine Airways of Juneau, Alaska. Marine Airways joined with Alaska Air Transport in 1940 to become Alaska-Coastal-Ellis Airlines. The aircraft continued to serve all these companies along the Alaska coast until sold to the Museum in May 1964 after 28 years of bush flying, a record that will be hard to equal. In 1945 it was re-engined with a more powerful Pratt & Whitney Wasp Jr. engine.

It is planned to restore the specimen with its original Wright J-6-9 engine and original-type Edo floats. Surprisingly, none of the record-breaking Bellancas are known to survive and this aircraft is the only known museum specimen of this outstanding aircraft family.

Boeing 247D

SPECIFICATIONS

Powerplant: Two 550 hp Pratt & Whitney Wasp S1H1-G

Wing span: 74 ft (22.6 m)

Length: 51 ft 7 in (15.7 m)

Height: 12 ft 6 in (3.8 m)

Wing area: 836 sq ft (77.7 m²)

Weight, empty: 8,940 lb (4055 kg)

Weight, gross: 13,650 lb (6192 kg)

Speed, max.: 200 mph (322km/h)

Speed, cruising: 189 mph (304 km/h)

Climb: 1,150 ft/min (350 m/min)

Service ceiling: 25,400 ft (7740 m)

February 8, 1933, was a significant day in the history of air transport, for on that day the Boeing 247 prototype, the first modern transport, made its initial flight. It entered scheduled service with United Air Lines the following May, cutting the travelling time between New York and San Francisco from 27 hours to 19 1/2 hours, a reduction of almost 30 per cent. It was not just the Boeing's increased speed that made it significant; it also established many "firsts" in safety and comfort, including the following: the first transport to climb with full load on one engine; the first with supercharged engines; the first with control trim tabs; and the first with cabin air conditioning. The introduction of the Boeing 247 earned William E. Boeing the 1934 Daniel Guggenheim Medal for pioneering achievement in aircraft manufacture and air transport.

The 247 had almost all the basic design features that became standard on later transport aircraft. It was an all-metal low-wing monoplane with engines in the wing leading edge and retractable undercarriage. However, it did not have wing flaps, and initially it did not have controllable-pitch propellers. The improved version, the 247D, had more powerful engines, controllable-pitch propellers and other refinements, and most 247s were brought up to 247D standards. Altogether 75 Boeing 247s were made.

The Boeing 247 was used initially mainly by United Air Lines, National Air Transport, Western Air Express and Pennsylvania Central Airlines. The type entered Canada in 1940 when the RCAF bought eight, which were used for training and transport duties until 1942. Five then went to Canadian Pacific Airlines and its subsidiaries, Quebec Airways and Yukon Southern Air Transport, and one went to Maritime Central Airways; all later returned to the United States.

The Museum's specimen was completed in the spring of 1934 as a

247 and was converted to a 247D in July 1935. It served with all of the US airlines noted above before joining the RCAF as serial 7638 in 1940. In June 1942 it became CF-BVX of CPA and was used in western Canada until sold in the US in 1945. There it served several owners until it returned to Canada in 1959 as CF-JRQ with the California Standard Oil Co. of Calgary. It continued in their service in western Canada until donated to the Museum on February 11, 1967.

The Museum's specimen is one of four examples of this significant transport aircraft preserved in museums, the others being two in the US and one in Britain.

*

Bristol 149 Bolingbroke IVT

SPECIFICATIONS

Powerplant: Two 920 hp Bristol Mercury XV.
Wing span: 56 ft 4 in (17.2 m)
Length: 42 ft 9 in (13 m)
Height: 9 ft 10 in (3 m)
Wing area: 469 sq ft (43.6 m²)
Weight, empty: 8,963 lb (4065 kg)
Weight, gross: 14,500 lb (6576 kg)
Speed, max.: 262 mph (422 km/h)
Climb: 10,000 ft (3048 m) in 6.6 min
Service ceiling: 28,400 ft (8660 m)

In late 1936, the RCAF sought a general reconnaissance aircraft and, in 1937, selected the Bolingbroke, an improved version of the Blenheim being developed for the RAF. The RAF then dropped the development of the Bolingbroke, which continued at the RCAF's request. The Blenheim from which the design started was itself a development of a civil type, the Bristol 142 *Britain First* sponsored by Lord Rothermere, which first flew in September 1935.

112

The British-built Bolingbroke prototype was first flown in September 1937, and the following November the RCAF placed an order with Fairchild Aircraft Ltd. at Longueuil, Quebec, for the type. Three versions of the basic design were produced there during the years 1939 to 1943. These versions included a seaplane made as a prototype only, and there were also minor variants of the production versions. The most numerous version was the Mk.IV, of which 608 were built in various sub-types.

The Bolingbroke was first introduced into the RCAF late in 1939 and in the early war years it served operationally on the east and west coasts of Canada and in Alaska. In the early war years there was little or no enemy activity in these areas so the type did not have to prove itself operationally. The Bolingbroke was developed into a bombing and gunnery trainer and target tug and in these capacities it served in the bombing and gunnery schools of the British Commonwealth Air Training Plan operated by the RCAF.

Following the war, Bolingbrokes were sold as war surplus and many survived in private hands until fairly recent years. No fewer than 13 are known to have found their way into museum collections in Belgium, Canada, England, Scotland and the United States, but it is believed that only the National Aviation Museum's specimen is restored and on display.

The Museum's specimen was accepted by the RCAF in June 1942 and was briefly on strength of No.1 Training Command at Toronto before going to No.2 Training Command at Winnipeg in July. It served at 4 TC until stored in June 1944 and was put up for disposal in September 1945. It was bought by George Maude, who stored it on Saltspring Island, British Columbia, until he donated it to the RCAF in 1964. The RCAF restored it in the markings of Bolingbroke IV 9025 of No.8 Squadron. The squadron had used Bolingbroke IVs operationally on both coasts from 1941 to 1943.

Canadair CL-28 Argus 2

In 1952 the RCAF needed a new aircraft type for anti-submarine patrol and maritime reconnaissance. It decided that the best and most economical answer to their requirement would be to modify the Bristol Type 175 Britannia rather than obtain a completely new design. Canadair Ltd. was given the contract to make the design changes and construct the aircraft.

The original design of the wings and tail surfaces was used, with wing modifications to carry armament, but a completely new fuselage

Powerplant: Four 3,700 hp Wright
R-3370 TC981 EA-1
Wing span: 142 ft 3 1/2 in (43.4 m)
Length: 128 ft 3 in (39.1 m)
Height: 37 ft 8-1/2 in (11.5 m)
Wing area: 2,075 sq ft (192.8 m²)
Weight, empty: 81,000 lb (36 742 kg)
Weight, gross: 148,000 lb (67 133 kg)
Speed, max.: 288 mph (464 km/h)
Speed, cruising: 207 mph (333 km/h)
Climb: 900 ft/min (274 km/min)
Service ceiling: 24,200 ft (7380 m)

was required and four Wright Turbo Compound engines of 3,700 hp each were installed to give the high power at altitude with maximum fuel economy needed to achieve its approximately 4,500-mile (7420-km) range and 26 1/2-hour endurance. Two bomb bays were provided to carry up to 8,000 lb (3630 kg) of weapons, which included active or passive torpedoes, bombs, depth charges, mines, practice bombs and photo flash bombs. Strong points were provided in the wings to mount missiles, rockets or other external stores as required.

The Argus was fitted with extensive electronic equipment, including high-definition medium-range search radar, electronic counter measures equipment, explosive echo-ranging equipment, magnetic anomaly detection equipment and sonobuoys. Also fitted was a high-intensity searchlight, search and rescue and homing equipment.

The Argus prototype, RCAF 20710, was first flown on March 28, 1957, at Cartierville, Quebec, and the type entered RCAF service under the RCAF designation the following May. The original version, the Argus 1, was followed about 17 months later by the Argus 2, which differed only in improved radar with smaller radome and other electronic equipment. Thirteen Argus 1s and 20 Argus 2s were delivered by the end of the manufacturing program in November 1960.

The Argus aircraft were used by three RCAF squadrons, Nos. 404 and 405 squadrons at Greenwood, Nova Scotia, and No.415 Squadron at Summerside, Prince Edward Island. In service the Argus performed well and its range and powerful electronic equipment made it most effective in its designed role until it was superseded by the Lockheed Aurora in 1981.

The Museum's specimen, Argus 2 RCAF 10742, was the last Argus made and served with 415 Squadron until its retirement. It was officially handed over to the Museum in a brief ceremony on May 17, 1982. It is the largest aircraft in the collection by a considerable margin.

Canadair CL-84-1 Dynavert

SPECIFICATIONS

Powerplant: Two 1,500 hp
Lycoming T.53

Wing span: 34 ft 8 in (10.6 m)

Length, max.: 53 ft 7-1/2 in
(16.3 m)

Height, wing at 90°: 17 ft 1-1/2 in
(5.2 m)

Wing area: 233.3 sq ft (21.7 m²)

Weight, empty: 8,775 lb (3980 kg)

Weight, gross, VTOL: 12,600 lb
(5714 kg)

Weight, gross, STOL: 14,500 lb
(6577 kg)

Speed, max., VTOL: 321 mph
(517 km/h)

Speed, cruising: 309 mph
(497 km/h)

Climb, VTOL: 4,200 ft/min
(1280 m/min)

One of the most interesting projects undertaken by the Canadian aircraft industry was the successful development of a V/STOL aircraft by Canadair Ltd. Designed to meet a NATO specification, it started as a private venture by Canadair and then was funded by the Defence Research Board and the Department of Defence Production. The requirement was for an aircraft which could take off vertically like a helicopter and also operate as a STOL aircraft, and which could be used for transport, ground support, reconnaissance and search and rescue operations.

After seven years of study, a prototype was built in just over a year. It featured a wing that could be tilted through 100° complete with twin powerplants swinging large-diameter propellers or rotors. The first prototype, the CL-84, was first flown in the hover mode on May 7, 1965, and the first transitions from hover to forward flight and back again were carried out on January 17, 1966. After completing 145 hours of flying, the CL-84 was lost in an accident without loss of life in September 1966, due, it was believed, to a failure of the differential propeller control system.

The success of the CL-84 led to the ordering of three improved versions, the CL-84-1s, which were generally similar to the CL-84 but more powerful and with control systems improved in details learned from the CL-84 trials. The first of these machines was flown on February 19, 1970.

The first two CL-84-1s were subjected to extensive tests by Canadair, the US Navy and US Marine Corps with successful results. The first CL-84-1 was lost in Chesapeake Bay following a gearbox failure, but the two US Navy and Marine Corps pilots ejected safely. The second CL-84-1, RCAF 8402, was also the subject of many successful trials as noted below. However, the type was not ordered into production.

The Museum's specimen, RCAF 8402, the third prototype and second CL-84-1, first flew in September 1972, then was taken on check flights by Canadian, American and British pilots before going to the USN Test Center at Patuxent River, Maryland, in December. During the next six months various instrument tests were made with it and on April 5, 1973, the world's first transition on instruments from conventional to hovering flight was made on it. Various trials were carried out on the carrier USS *Guam* late in 1973 and still more on board the USS *Guadalcanal* in March 1974, as well as night flying evaluation in June 1974. Altogether it made 196 flights totalling 169.1 flying hours. It was donated to the Museum in April 1984 with components being delivered from April through June.

Cessna T-50 Crane I

The Cessna T-50 was designed as a small, low-cost, utility twin-engined monoplane for civil use to seat five people including the pilot. The prototype was first flown in March 1939 and type approval was received in 1940, with the first civil deliveries made the same year. In 1940 the RCAF placed an order for the type, which they intended to use as a trainer to supplement the Avro Anson. This was the first large order ever received by the Cessna Aircraft Co. and led to its expansion; then the US Army Air Force ordered the T-50 in various versions in quantity. In American service the type was known as the Bobcat and given military designations AT-8, AT-17 and UC-78.

SPECIFICATIONS

Powerplant: Two 225 hp
                Jacobs L-4MB
Wing span: 41 ft 11 in (12.8 m)
Length: 32 ft 9 in (10 m)
Height: 9 ft 4 in (2.8 m)
Wing area: 295 sq ft (27.4 m²)
Weight, empty: 3,500 lb (1587 kg)
Weight, gross: 5,100 lb (2313 kg)
Speed, max.: 185 mph (298 km/h)
Speed, cruising: 165 mph (266 km/h)
Climb: 1,200 ft/min (365 m/min)
Service ceiling: 18,000 ft (5490 m)

The RCAF took delivery of 826 Cranes, of which 744 were Crane Is and 82 were the more powerful Crane II. The first Crane was taken on strength in January 1941. They all served in western Canada with the following service flying training schools: No.3 at Calgary, No.4 at Saskatoon, No.11 at Yorkton, Saskatchewan, No.12 at Brandon, Manitoba, and No.15 at Claresholm, Alberta. The aircraft were used to give twin-engine training to graduates of the elementary flying training schools before they received their wings and proceeded to service flying training. The type continued in RCAF service until September 1947.

Cranes were sold as war surplus for civil use in Canada and the United States, where they took up their original design role as light transports but few remain in use today. There are two examples of the type registered in Canada in 1986, one of them to the Canadian Warplane Heritage in Hamilton.

The Museum's specimen was taken on strength by the RCAF in January 1942 and served initially with 15 Service Flying Training School, then went on to Prairie Airways Ltd. of Moose Jaw, Saskatchewan, at the end of June 1943, apparently on loan for some unknown purpose. It then went to No.12 SFTS at Brandon, Manitoba, until it was stored in 1946. It was shipped to the RCAF's historic aircraft collection at Rockcliffe in 1964.

Consolidated *PBY-5A* *Canso A*

SPECIFICATIONS

Powerplant: Two 1,200 hp. Pratt & Whitney Twin Wasp R-1830-92

Wing span: 102 ft (31 m)

Length: 63 ft 10-1/2 in (19.5 m)

Height: 20 ft 2 in (6.1 m)

Wing area: 1,400 sq ft (130 m²)

Weight, empty: 20,910 lb (9485 kg)

Weight, gross: 33,975 lb (15 411 kg)

Speed, max.: 179 mph at 7,000 ft (288 km/h at 2133 m)

Speed, cruising: 117 mph at 7,000 ft (188 km/h at 2133 m)

Climb: 10,000 ft (3050 m) in 19.3 min

Service ceiling: 20,000 ft (6100 m)

The prototype of the PBY flying boat first flew in March 1933 as the XP3Y-1 and went into service with the US Navy as the PBY-1. By 1939 the production version was the PBY-5 and an amphibious version, the PBY-5A, had just been introduced. During World War II the type was built in Canada, the USA and the USSR, and in all 3,431 were built of the different versions, more than any other flying boat, which is a record that will endure. It was used during the war by the RAAF, RAF, RCAF, RNZAF, USAAF, US Navy and Soviet Navy.

In 1939 the RCAF selected the type as the Supermarine Stranraer's successor and in 1940 placed production orders with Boeing Aircraft of Canada Ltd. at Vancouver and Canadian Vickers Ltd. at Montreal. Both firms went on to produce the type not only for the RCAF but also for Canada's allies. The long range and endurance of the PBY were badly needed to counter the German submarine menace in the Atlantic. While the RAF and other Commonwealth air forces used the name Catalina for the type, because of equipment differences the RCAF named the type Canso for the flying boats and Canso A for the amphibians.

The first Cansos joined the RCAF in June 1941 and the first Canso As the following August, and they were widely used in the Battle of the Atlantic as well as on the Pacific coast. A batch of Boeing Canada-made Canso As went to 162 Squadron, which distinguished itself by accounting for all six enemy submarines destroyed by Canso As in the RCAF. Included in this group was the sinking of the U-1225 by F/L D.E. Hornell and crew; in the aftermath Hornell lost his life and he was awarded a posthumous Victoria Cross for his actions.

After the war Canso As continued to serve with the RCAF until retired in November 1962. Their postwar role was primarily photo reconnaissance and search and rescue operations. Other Canadian Canso As went on to serve with the Royal Danish and Royal Swedish Air Forces. Many Canso As were converted for civil flying in passenger service, water bombing, freight service and aerial surveying, and some still remain in use.

The Museum's specimen was made by Canadian Vickers in June 1944 and served initially with Eastern Air Command. After a brief period in storage, it served from coast to coast on search and rescue duties until retired in 1962. It was then refinished by the RCAF in the markings of F/L Hornell's aircraft of 162 Squadron. Unfortunately it has had to be stored outside and the finish has weathered.

Consolidated B-24L Liberator G.R. VIII

SPECIFICATIONS

Powerplant: Four 1,200 hp Pratt & Whitney Twin Wasp R-1830-65
Wing span: 110 ft (33.5 m)
Length: 67 ft 2 in (20.5 m)
Height: 18 ft (5.5 m)
Wing area: 1,048 sq ft (97.4 m²)
Weight, empty: 36,500 lb (16 550 kg)
Weight, gross: 64,500 lb (29 250 kg)
Speed, max.: 300 mph (483 km/h)
Speed, cruising: 215 mph (346 km/h)
Climb: 20,000 ft (6100 m) in 5 min
Service ceiling: 28,000 ft (8530 m)

The Liberator bomber was a remarkable aircraft in a number of aspects. The design was only started at the beginning of 1939 and yet the prototype was first flown on December 29, 1939 – an outstanding achievement. It was placed in production at two Consolidated plants and three other plants, including the Ford plant at Willow Run, Michigan, which eventually produced a Liberator an hour. A total of 18,481 Liberators was produced and delivered to Allied air forces, more than any other American-made aircraft. They were operated during the war by the RAAF, RAF, RCAF, USAAF and USN.

The outstanding feature of the Liberator was its range of 2,100 miles (3380 km). Its tricycle undercarriage was unusual for the time and the first to be fitted to a large bomber. It normally carried a crew of eight to ten and a bomb load of up to 8,000 lb (3630 kg). Defensive armament consisted of ten 0.50 guns.

Its usual role was bombing operations against Germany, occupied Europe and Japan, in which it proved most effective. Less publicized was its role in the war against the submarine, where its range and speed were put to good use; in one instance five RAF Liberator squadrons sank six submarines in seven days. There seems little doubt that the Liberator was the single most effective aircraft weapon against the submarine.

The RCAF operated four Liberator anti-submarine squadrons, Nos. 10, 11, 422 and 423; one heavy transport squadron, No.168; and one bomber squadron that was just re-equipped with Liberators for the war against Japan when it ended. No.10 Squadron sank two submarines with Liberators, while No.11 carried out eight submarine attacks. Nos. 422 and 423 were re-equipped too late to use Liberators operationally. One RCAF squadron, No.412, used Liberators for transport duties in the postwar period.

The Museum's specimen was made by the Ford Motor Co. in 1944 as USAAF 44-50154 and went to the RAF on the lease-lend program. After being abandoned in India, it was refurbished and taken on strength by the Indian Air Force in 1949 as HE773 shortly after India was granted self-government. It served with No.6 Squadron, IAF, until June 1968, when it was given to the RCAF in appreciation for its earlier donation of a Westland Lysander. It was flown to Canada and refinished as an anti-submarine aircraft of Eastern Air Command by 6 RD, Trenton, Ontario, before being placed on display.

Curtiss HS-2L

The Curtiss Aeroplane and Motor Co. decided to develop a more powerful flying boat in 1917. The prototype, the HS-1, powered with a 200 hp Curtiss engine, first flew in the early summer of that year. The US Navy became interested in the aircraft and had the new Liberty engine tested in the prototype in October. A slightly modified Liberty-powered model was then placed in production as the HS-1L. Its wing span was then increased to enlarge its carrying capacity and this version was designated the HS-2L. A total of 1,121 HS-1Ls and HS-2Ls was made by six manufacturers.

During World War I the type flew anti-submarine patrols from bases in France, the United States and Canada. It first appeared in Canada when the USN flew coastal patrols from bases at Sydney and Dartmouth, Nova Scotia, to protect convoys sailing from Halifax. Twelve HS-2Ls at those bases were donated to Canada in 1919; two of these were immediately loaned for civil use and the others started flying with the Air Board the following year. These aircraft were soon

SPECIFICATIONS

Powerplant: One 360 hp Liberty

Wing span, upper: 74 ft-19/32 in
(22.6 m)

Wing span, lower: 64 ft 1-21/32 in
(19.55 m)

Length: 39 ft (11.9 m)

Height: 18 ft 9 1/4 in (5.7 m)

Wing area: 803 sq ft (74.6 m²)

Weight, empty: 4,300 lb (1950 kg)

Weight, gross: 6,432 lb (2918 kg)

Speed, max.: 85 mph (137 km/h)

Speed, cruising: 60-65 mph
(97-105 km/h)

Climb: 1,800 ft (549 m) in 10 min

Ceiling: 9,200 ft (2800 m)

joined by many others purchased as war surplus. The type became not only the first Canadian bushplane but also the predominant bushplane until 1926 or 1927, and the last was only retired from civil use in 1932.

The HS-2L flew our first forestry patrols, made our first aerial timber survey in 1919, staked the first mining claim with the use of aircraft in July 1920 and in 1924 established both the first scheduled air service and first regular air mail service in Canada, serving the Quebec goldfields (now Rouyn-Noranda). The RCAF and its predecessor, the Air Board, operated the largest number of HS-2Ls in Canada, but it was Laurentide Air Service Ltd. and its predecessors that recorded all the "firsts" noted above and that established the traditions of Canadian bush flying using the type. The Ontario Provincial Air Service, which started operations in 1924, had a fleet of 19 and was the last to use the HS-2L in Canada, in 1932.

Slow, unwieldy and uneconomical by comparison with its high-wing monoplane successors, nevertheless the HS-2L pioneered Canadian bush flying and started the opening up of vast areas of the north. Canada owes much to this early workhorse of the bush.

The history of the Museum's specimen is told more fully elsewhere in this book, but, briefly, it incorporates parts from our first bush aircraft, G-CAAC of Laurentide, G-CAOS of the Ontario Provincial Air Service, and an American machine of Pacific Marine Airways.

Curtiss JN-4 (Can.) "Canuck"

SPECIFICATIONS

Powerplant: One 90 hp Curtiss OX-5

Wing span, upper: 43 ft 7 3/8 in
(13.29 m)

Wing span, lower: 36 ft 8-5/16 in
(11.18 m)

Length: 27 ft 2 1/2 in (8.29 m)

Height: 9 ft 11 in (3.02 m) approx

Wing area: 360.6 sq ft (33.5 m²)

Weight, empty: 1,390 lb (630 kg)

Weight, gross: 1,930 lb (875 kg)

Speed, max.: 74 mph (119 km/h)

Speed, cruising: 60 mph (97 km/h)

Climb: 2,500 ft (762 m) in 10 min

Ceiling: 11,000 ft (3350 m)

RIGHT
*The right side of the "**Canuck**" was left open so visitors can see its construction and other internal details.*
(KMM)

FAR RIGHT
*The instrument panel of the "**Canuck.**" Top row: two instrument lights and light switches. Centre row: airspeed indicator, Canadian Aeroplanes Ltd. construction no. plate, altimeter, tachometer. Bottom: oil pressure gauge. Note throttle lever on right, opposite to modern practice. Compass was mounted on shelf beneath the panel but is not installed here.*
(NMC)

The JN-4 (Canadian) was a modification of the American Curtiss JN-3, which had been made by Curtiss Aeroplanes and Motors Ltd. at Toronto. The changes were made by Canadian Aeroplanes Ltd. at Toronto in December 1916 to suit the requirements of the RFC (Canada), which was to establish a Canadian training program in 1917. The prototype first flew early in January 1917 and went into large-scale production initially for the RFC (Canada) and shortly for the US Air Service also. In all, Canadian Aeroplanes made about 1,210 of the type, plus many spares, and in the postwar years about 44 were made by Ericson Aircraft at Toronto and J.V. Elliot at Hamilton.

As no suffix had been given the type to distinguish it from the various American JN-4 versions of the Jenny family, it became known as the JN-4 (Canadian) and inevitably as the "Canuck" the only member of the Jenny family to have its own name. It went into military use in Canada in February 1917 and did almost all the training in Canada during World War I. When introduced into the US Air Service, it contributed more to the American training program than is usually realized as it was available in quantity early in their program while American types were delayed getting into service. In the postwar period it saw widespread civil use in Canada and the United States, flown by many barnstormers and preferred by most, and it only disappeared from the skies about 1930.

During this long period of service, the Canuck established a secure niche in Canadian aviation history by recording more "firsts" than any other type. It was the first type to be made in Canada in large quantities and the first to be exported in large quantities. It performed the first military flying in Canada and the first ski flying in the winter of 1917-18. It flew the first Canadian air mail in June 1918 between Montreal and Toronto and made the first aerial survey in Labrador in July-August 1919, which may even be a world's first. Finally it was the usual equipment of the Canadian barnstormer of the postwar era, so it was normally a Canuck that first appeared in many towns across the country to give Canadians their first close look at an aircraft and their first chance to fly.

The Museum's specimen was originally supplied to the USAS as serial 39158 and was almost certainly based at Love Field, Dallas. It entered civil use in the postwar period but the details are not known. It was then owned by George Reese of Naples, New York, who sold it, probably in 1925 or 1926, to Edward Faulkner of Honeoye Falls, New York. It was flown from Rochester, registered civilly in 1927 as 111, and stored at Honeoye Falls in 1932 until sold to the Museum in 1962.

Curtiss Seagull

The Seagull was a civil development of the Curtiss MF flying boat introduced in early 1918 as a trainer by the USN to replace its aging Curtiss F flying boats. The MF introduced a considerable change in hull lines and construction over earlier Curtiss flying boats, with the hull being formed of mahogany plywood veneer over a wood frame. The standard powerplant of the MF was a 100 hp Curtiss OXX-2 or OXX-6. The type was produced by both Curtiss and the Naval Aircraft Factory, and about 122 were built.

The Seagull appeared early in 1919. It remained basically the same as the MF, but the hull could now accommodate three people in place of

SPECIFICATIONS

Powerplant: One 160 hp Curtiss C-6
Wing span, upper: 49 ft 9 in (15.2 m)
Wing span, lower: 38 ft 7-5/32 in
                (11.76 m)
Length: 28 ft 10 3/16 in (8.8 m)
Height: 11 ft 9 7/32 in (3.6 m)
Wing area: 401 sq ft (37.3 m²)
Weight, empty: 1,957 lb (888 kg)
Weight, gross: 2,726 lb (1237 kg)
Speed, max.: 76 mph (122 km/h)
Speed, cruising: 60 mph (97 km/h)
Climb: 3,000 ft (914 m) in 10 min
Service ceiling: 5,900 ft (1800 m)

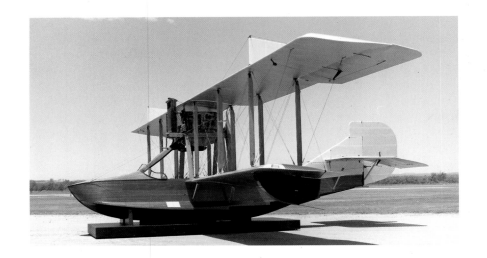

two and a more powerful engine was installed, initially a 150 hp Curtiss K-6 and later a 160 hp Curtiss C-6.

The Seagull was judged the best of the smaller single-engined flying boats of its time, and while it suffered in competing with the war-surplus and consequently low-priced MF flying boats, its increased power and accommodation made it the choice of a number of fixed-base operators. Chaplin Aircraft Corp. operated an air service with Seagulls to Catalina Island off the California coast. Several pioneering flights with the type brought it publicity, including J.L. Larsen's tour of the Baltic Sea, piloted by Carl Batts, and a long flight by Capts. S.E. Parker and G.T. Wilcox from New York to Montreal, then along the Great Lakes and down the Mississippi to New Orleans, barnstorming en route.

The Seagull was somewhat small for the heavy work usually involved in Canadian bush flying, but nevertheless several Seagulls, or their modified ancestors, the MFs, were used. The first was imported by the Laurentide Co. in 1920 and another had an interesting career carrying fishermen with the Grey Rocks Air Service, but probably the most interesting Canadian Seagull was a highly modified version with new wings that served Fairchild Aerial Surveys (of Canada) beginning in 1925.

Likely the best known of the Seagull's ventures was carried out by the Museum's specimen. It was used in an aerial survey of the upper Amazon River in South America in the region of the equator in 1924-25. The machine was then given to the South Kensington Science Museum in London, England, and came to the National Aviation Museum in 1968 as a result of an exchange. It was restored by 1974 and has been on display ever since.

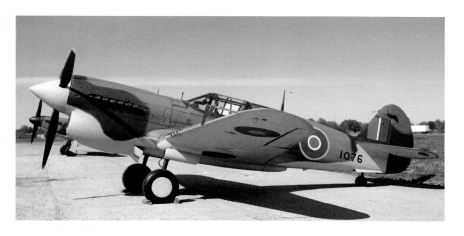

Curtiss P-40D Kittyhawk I

SPECIFICATIONS

Powerplant: One 1,150 hp Allison V-1710-39

Wing span: 37 ft 3-1/2 in (11.4 m)

Length: 31 ft 8-1/2 in (9.7 m)

Height: 10 ft 8 in (3.3 m)

Wing area: 236 sq ft (21.9 m²)

Weight, empty: 5,922 lb (2686 kg)

Weight, gross: 8,515 lb (3862 kg)

Speed, max.: 334 mph (538 km/h)

Speed, cruising: 308 mph (496 km/h)

Climb: 15,000 ft (4570 m) in 6 min 15 sec

Service ceiling: 29,100 ft (8870 m)

The Curtiss P-40 was the last production version of the well known Curtiss Hawk series of fighters which started in 1923. Altogether 13,738 of the various P-40 variants were built and some of them were given the names Tomahawk, Kittyhawk and Warhawk. The RCAF operated all three of the named variants of the type.

The P-40 was built to a US Army Air Corps specification for a low-altitude fighter in the 1930s. This differed from contemporary European specifications, which were for high-altitude fighters with considerably heavier armament and defensive provisions. The result was that when the prototype first flew in October 1938 it was well behind its contemporaries in Europe. Many improved versions of the P-40 appeared which helped make up the deficiencies, but the type was never able to catch up with its contemporaries, even when one version, the P-40L, had the Packard-built Rolls-Royce Merlin fitted. Nevertheless P-40s in the hands of the RAF and fitted out to carry bombs proved most effective in low-level attacks against Rommel's armies in North Africa.

The Kittyhawk I was the first of the P-40 series used by the RCAF. Its service began in November 1941 and it was joined later by Kittyhawk IAs, IIIs and IVs until 143 had been taken on strength. Armament was six 0.50 guns. Kittyhawks served in the following RCAF squadrons: Nos. 14, 111, 118, 130, 132, 133 and 163. All these squadrons were employed on defensive patrols in Canada and Alaska but only 111 and 118 participated in any offensive action. On September 25, 1942, S/L K.A. Boomer of 111 Squadron shot down a Japanese Nakajima A6M2-N (Ruff) fighter in the only RCAF aerial combat in the North American theatre of war. On January 16, 1942, two Kittyhawks from 118 Squadron, stationed at Dartmouth, Nova Scotia, surprised a German submarine on the surface near Halifax and attacked it with gunfire. All Kittyhawks were retired from the RCAF in 1946.

The Museum's specimen was taken on strength by the RCAF on March 23, 1942, originally aircraft AL135 on loan from the RAF. It was issued to 132 Squadron, forming at Rockcliffe, Ontario, and it then moved to British Columbia, where it was stationed successively at Sea Island, Patricia Bay, Tofino and Boundary Bay. The aircraft then went to No.3 Training Command at Montreal in January 1944 and to No.2 Training Command at Winnipeg in June 1945. It was then stored at Vulcan, Alberta, until it was brought to Rockcliffe in 1964 and placed on display.

＊

Czerwinski/Shenstone Harbinger

SPECIFICATIONS

Wing span: 60 ft (18.3 m)
Length: 23 ft 9 in (7.2 m)
Wing area: 240 sq ft (22.3 m²)
Weight, empty: 670 lb (304 kg)
Weight, gross: 1,030 lb (467 kg)
Speed, max.: 130 mph (209 km/h)
Min. sinking speed at 42.5 mph (68 km/h): 2 ft 6-3/4 in/sec (78 cm/sec)

The Harbinger was designed at Toronto by Waclaw Czerwinski and Beverly S. Shenstone to compete in a 1947 design competition for a two-seat sailplane sponsored by the British Gliding Association. Two-seat sailplanes were considered a difficult design challenge as there were problems in achieving good vision for both occupants. Six designs were submitted and the Harbinger only placed fifth, but in spite of this it was the only one actually to be built and flown. It was named Harbinger by Jack W. Ames, who felt that it promised to begin a new era in two-seat sailplanes.

The first of the type was built in England by Fred Colman of Rugby and was first flown in 1958. As a result of experience with this machine, the wing sweep was altered from 4° forward to 1° on the Canadian machine to correct a centre of gravity problem.

Construction of the Canadian Harbinger was started in Toronto in 1949 by Jack Ames and Henry Dow. Work ceased in 1951 when Ames' job relocated him and the parts were moved to London, Ontario, where Al Pow carried out further work until about 1953. The parts were bought in 1956 by A.N. Le Cheminant and R. Noonan and taken to Gimli, Manitoba, for completion by Le Cheminant, but he was moved to Ottawa in 1958 and not until 1967 did construction continue.

Finally assembly began at Pendleton, Ontario, just east of Ottawa, in 1975 and altogether over 30 flights were made, with the longest being over three hours in October 1976. While Le Cheminant, who did almost all the flying, said "the Harbinger was truly a dream and ahead of the state of the art when designed," there were a number of changes needed and he decided to discontinue development of the type. He donated the Harbinger to the National Aviation Museum in September 1977.

De Havilland D.H.60X Moth

SPECIFICATIONS

Powerplant: One 85 A.D.C. Cirrus II

Wing span, upper and lower: 30 ft (9.1 m)

Length: 23 ft 8-1/2 in (7.2 m)

Height: 8 ft 9-1/2 in (2.7 m)

Wing area: 242 sq ft (22.5 m²)

Weight, empty: 890 lb (404 kg)

Weight, gross: 1,550 lb (703 kg)

Speed, max.: 95 mph (152 km/h)

Speed, cruising: 85 mph (137 km/h)

Climb: 650 ft/min (198 m/min)

Ceiling: 17,000 ft (5180 m)

The prototype Moth first flew in England in February 1925 piloted by its designer, Geoffrey de Havilland (later Sir Geoffrey). It was immediately recognized as an outstanding type and put into production. It was made with several different engines, the most popular being the A.D.C. Cirrus and D.H. Gipsy engines of several Mk. numbers. Over its production life, modifications were made to the airframe, most significantly the introduction, at the request of Canadian authorities, of a metal tube fuselage in place of the original wooden one. The Moth was widely accepted internationally, was made under licence in Australia, Finland, France, Norway and the United

A

B

C

States, and became probably the most widely used type of its time. Production appears to have totalled 1,703 in England and 270 by the licensees.

The Moth was used as a trainer by many air forces and flying clubs throughout the British Empire and by private owners in many countries. Many long-distance flights were made in Moths, including Amy Johnson's flight from England to Australia in 1929, and Francis Chichester's flight from England to Australia in 1929 and across the Tasman Sea in 1930. Moths were also used in both the Arctic and Antarctic regions. Unfortunately an Atlantic flight in 1928 failed and its pilot, Lt. Cdr. H.C. MacDonald, was lost at sea.

The first Moth came to Canada in 1927 and was used by the Hudson Strait Expedition. It and some 26 others were shipped direct from England to Canadian customers until the de Havilland Aircraft of Canada Ltd. was established in April 1928 to assemble and service Moths in Canada. The type became the standard trainer of the RCAF and the flying clubs, and the majority of Canadian Moths served in this capacity. They were also used by private owners and, to a lesser extent, in bush flying, with Dominion Explorers, General Airways, Prospectors Airways and others. In 1932 J.R. Hébert of Montreal shipped his Moth, CF-ADC, to England and flew in stages to Australia. In the late 1920s and during the 1930s there were more Moths in Canada than any other type; the majority of Canadians trained on them, and there were few Canadian pilots of the time who had not flown a Moth.

The Museum's specimen, a Cirrus Moth, went into service in 1928 with International Airways of Canada at Hamilton and Toronto. Canadian Airways took over International Airways in November 1930 and G-CAUA was sent to Moncton, New Brunswick, and sold there in June 1933. It remained in the hands of private owners, including C.F. Burke, until 1942, when it was bought by Burke again and stored at Charlottetown, Prince Edward Island. He gave it to the Museum in May 1962. It was restored by old de Havilland Canada employees with the assistance of the company and the restored aircraft was turned over to the Museum on May 10, 1963.

A

The Moth with wings folded for convenient and economical storage and transport. (KMM)

B

The Moth's cockpit. Instruments are, left to right, oil temperature gauge,

tachometer, altimeter, airspeed indicator, oil pressure gauge, partly hidden in photo. The compass is mounted ahead of the control column on the floor. Throttle lever is on the left and below it can just be seen the knob of the trim lever. (KMM)

C

A close view of the Moth's tail surfaces. Note the characteristic shape of the rudder. First used on the D.H. 3 of 1916, it was used on almost every British de Havilland design thereafter. (KMM)

De Havilland D.H.80A Puss Moth

SPECIFICATIONS

Powerplant: One 130 hp
D.H. Gipsy Major
Wing span: 36 ft 9 in (11.2 m)
Length: 25 ft (7.6 m)
Height: 6 ft 10 in (2.1 m)
Wing area: 222 sq ft (20.6 m²)
Weight, empty: 1,265 lb (574 kg)
Weight: 2,050 lb (930 kg)
Speed, max.: 128 mph (206 km/h)
Speed, cruising: 105 mph (169 km/h)
Climb: 660 ft/min (200 m/min)
Ceiling: 17,500 ft (5334 m)

The wide acceptance of the D.H.60 Moth pointed to the need for a faster aircraft with an enclosed cabin for more speed and comfort. The reliable D.H. Gipsy engine was inverted and installed in a high-wing monoplane with all-wood structure. The prototype, the D.H.80, was first flown in September 1929. A second prototype, the D.H.80A Puss Moth, with a welded-steel-tube fuselage, was flown the following March. The type was essentially a two- but occasionally a three-seater, and an unusual feature was an air brake provided by turning the undercarriage strut 90°, which slowed the machine and steepened the glide.

The Puss Moth won immediate acceptance and was put into service by many private owners and fixed-base operators. A number of long-distance flights were made in the type, including ones from England to South Africa and Australia and Jim Mollison's Atlantic flight to New Brunswick. Early in the Puss Moth's career a series of fatal accidents provided a setback, and only after F/O A.L. James of the RCAF escaped from one at Rockcliffe was wing flutter determined to be the cause and the problem rectified.

The fourth Puss Moth made came to Canada, and the type was bought by the RCAF, private owners and commercial operators; the RCAF acquired 17 and 16 entered civil use. (Some machines were registered to both the RCAF and civil operators at different times.) One of the Canadian Puss Moths, CF-APK, became famous when H.J.L. "Bert" Hinkler, the well known Australian pilot, flew it in 1931 from Toronto to South America, then across the South Atlantic and north to England. CF-APK was the first Canadian aircraft to visit many of the countries he landed at. He lost his life in 1933 in 'APK when crossing the Alps en route to Australia. The type continued to give service until the present. Although no Canadian Puss Moth is now active, at least one is extant. The only appearance of the Puss Moth on floats was a very brief one by DHC's demonstrator.

The Museum's specimen had a long, varied career. It was originally bought by the US Navy as 8877 in 1931 for use by their London air attaché. In 1939 it was impressed as HM534 by the RAF, which it served until 1946. It then entered civil use as G-AHLO with various operators until bought in 1969 by Father John MacGillvray, who brought it to Canada. It was refinished in the registration and colours of a 1932 Puss Moth flown by Mrs. Louise Jenkins of Charlottetown, Prince Edward Island. On Father MacGillvray's retirement from the RCAF in 1976, it was sold to the Museum.

De Havilland D.H.82C2 Menasco Moth I

SPECIFICATIONS

Powerplant: One 125 hp Menasco D.4 Pirate

Wing span, upper and lower: 29 ft 4 in (8.9 m)

Length: 24 ft 2 in (7.4 m)

Height: 8 ft 9 1/2 in (2.7 m)

Wing area: 239 sq ft (22.2 m²)

Weight, empty: 1,229 lb (557 kg)

Weight, gross: 1,825 lb (829 kg)

Speed, max.: 96 mph (155 km/h)

Speed, cruising: 80 mph (129 km/h)

Climb: 600 ft/min (183 m/min)

Ceiling: 12,000 ft (3660 m)

The Menasco Moth is almost identical to its better known Tiger Moth parent except for the installation of the Menasco engine. The Tiger Moth was developed from the D.H.60 Moth, the main changes being the sweep-back of the wings to give better access to the front cockpit and the increased dihedral angle of the lower wings; the use of the inverted D.H. Gipsy engine was continued from the D.H.60GIII. The Tiger Moth prototype was first flown in England in October 1931.

The Tiger Moth was essentially a military trainer, with only small numbers being sold originally for civil use, but following World War II many military trainers came into civil hands. In addition to the RAF, for whom the type was developed, the Tiger Moth was used by the air forces of Australia, Brazil, Canada, Denmark, Iraq, New Zealand, Persia, Portugal, South Africa, Southern Rhodesia and Sweden. Its widespread use has made the type one of the best known trainers, like the Avro 504K of World War I.

The first Tiger Moth came to Canada as a civil demonstrator in 1935 and shortly the RCAF gave de Havilland Canada an order, which initiated aircraft manufacturing (as opposed to assembly) by DHC. The machines under the first contract were modified from the British design and known as D.H.82A (Can.). Still more modifications were specified on further orders and these machines were designated D.H.82C. Engines for the Tiger Moths were made in Britain and concern was felt about their continuing supply under wartime conditions, so Menasco-engined versions were built. Some were built as standard trainers, the DH.82C2 Menasco Moth Is, and others as wireless trainers, the D.H.82C4 Menasco Moth IIs. As the Menasco engine developed 125 hp, compared with the 140 hp of the D.H. Gipsy Major 1C, the Menasco Moth's performance suffered somewhat. The Tiger and Menasco Moths were identical in appearance except that the engine cooling air inlet was on the right on the Menasco and on the left on the Tiger Moth, and they had opposite-handed propellers.

Tiger and Menasco Moths, along with Fleet Finches, did all the elementary training in the British Commonwealth Air Training Plan until 1943, when they began to be replaced. Following the war, Tiger and Menasco Moths were sold on the civil market and were widely used, with some appearing on floats. Many still fly today and they have become collectors' items.

The Museum's specimen was built as a D.H.82C4 Menasco Moth II wireless trainer in 1941 and served with No.1 Wireless School at Winnipeg. In 1943 it was converted to a D.H.82C2 Menasco Moth I elementary trainer and served at No.5 Air Observer School at Winnipeg. It was stored in the postwar period and came to the RCAF historic aircraft collection at Rockcliffe in 1964.

Just before World War II, de Havilland proposed a clean, unarmed bomber relying entirely on its speed for protection. This idea was at first turned down by the British Air Ministry but, influenced by a single Air Member, the Air Ministry ordered a prototype in December 1939. The aircraft, now named the Mosquito, first flew on November 25, 1940, and demonstrated that it was the fastest operational machine, a distinction it retained for over two years. Large production orders were placed and the type was used as an unarmed bomber,

De Havilland D.H.98
Mosquito B.Mk.XX

131

Powerplant: Two 1,300 hp
        Packard Merlin 33

Wing span: 54 ft 2 in (16.5 m)

Length: 40 ft 4 in (12.3 m)

Height: 12 ft 3 1/2 in (3.7 m)

Wing area: 450 sq ft (41.8 m²)

Weight, gross: 21,980 lb (9970 kg)

Speed, max.: 369 mph (594 km/h)

Speed, economic cruising: 200 mph at
        15,000 ft
        (322 km/h at
        4570 m)

Climb: 2,200 ft/min (670 m/min)

Service ceiling: 34,500 ft (10 520 m)

photo-reconnaissance aircraft, armed fighter, trainer and special-purpose aircraft. One of its distinctive features was that its airframe was constructed entirely of wood.

Mosquitos were successful in all their roles. The bombers carried a 4,000-lb (1815-kg) bomb and their speed made them and the photographic Mosquitos difficult for enemy aircraft to intercept. The fighter, armed with four 20mm cannon and four 0.303 guns, proved exceedingly effective against enemy aircraft. Versions developed for specific purposes also proved their worth, including one armed with a 57mm quick-firing gun to attack shipping and pressurized versions for high-altitude reconnaissance and to intercept high-flying Junkers Ju 88Ps.

Orders were placed with de Havilland of Canada in August 1941 for Mosquitos and the Canadian prototype first flew on September 23, 1942. Canadian production included eight different variants, among them bombers, fighter-bombers, photo-reconnaissance machines for the USAF and trainers. 1,031 Canadian Mosquitos were accepted during the war and about another hundred were completed in the postwar period. About 205 Canadian Mosquitos were supplied to Nationalist China in 1947-48.

Canadian-built Mosquitos served with the RAF on all the usual Mosquito operations and the USAF used their photo-reconnaissance version over Europe from British bases. The RCAF Mosquitos were used at operational training units at Debert, New Brunswick, and Greenwood, Nova Scotia. The RCAF made little use of them in the postwar period and they were all disposed of by 1950. Several Canadian-built Mosquitos appeared on the Canadian civil register in postwar years; most were used for surveying but one was flown in the Bendix Trophy Race in 1948.

The Museum's specimen, a Mk.XX bomber, was the second variant built by DHC. It was accepted by the RAF on June 1, 1944, and transferred to the RCAF on June 12. It served with No.7 OTU at Debert, and was stored in 1945 until transferred to the RCAF's historic aircraft collection at Rockcliffe in 1964.

De Havilland D.H.100 Vampire 3

SPECIFICATIONS

Powerplant: One 3,100 lb (1405 kg) st de Havilland Goblin 2
Wing span: 40 ft (12.2 m)
Length: 30 ft 9 in (9.4 m)
Height: 8 ft 10 in (2.7 m)
Wing area: 226 sq ft (21 m²)
Weight, empty: 7,134 lb (3235 kg)
Weight, gross: 11,970 lb (5430 kg)
Speed, max.: 531 mph (855 km/h)
Climb: 4,350 ft/min (1325 m/min)
Ceiling: 43,500 ft (13 260 m)

On September 20, 1943, the de Havilland Vampire prototype became the third British jet aircraft type to fly, after the Gloster E.28/39 and the Gloster Meteor. The first production Vampire did not fly until April 1945 so the type was too late to take part in World War II.

The aircraft was of all-metal construction except for a cockpit section of balsa wood sandwich construction like the D.H. Mosquito. De Havilland also became the third British company to develop a jet engine, after Power Jets Ltd. (Whittle) and Metropolitan-Vickers Electrical Co. Their first production engine, the Goblin, powered the majority of Vampire models, with a few late Vampires being fitted with the more powerful Rolls-Royce Nene. Armament of the Vampire 3 consisted of four 20mm guns.

Being relatively economical and readily available in the postwar years, the Vampire was put into service by the following countries: Australia, Britain, Canada, Egypt, Finland, Italy, Iraq, Lebanon,

Norway, Sweden, Switzerland and Venezuela. Some Vampires were later passed on to still other countries; the type was made under licence in Australia, Italy and Switzerland and some were assembled in France from British components. Six Vampires of No.54 Squadron RAF made aviation history by flying the Atlantic in stages in July 1948, the first jet fighters to do so.

The Vampire first entered RCAF service in 1946, but this was only a trial Vampire 1 and it was not until 1948 that the air force's 85 Vampire 3s began to arrive. These were the first jet aircraft to enter RCAF squadron service. They were assigned to the following RCAF auxiliary squadrons: No.400, City of Toronto; No.401, City of Montreal; No.402, City of Winnipeg; No.438, City of Westmount; and No.442, City of Vancouver. Only one regular RCAF squadron was equipped with Vampires, No.421 at Chatham, New Brunswick. Some RAF Vampire 5s on loan from the RAF in 1952 served with No.421 Squadron in England in 1951. Vampires were withdrawn from use by the RCAF in 1958.

The Museum's specimen was taken on strength at St. Hubert, Quebec, in August 1948 and served with No.1 Fighter OTU there until November 1949. It served with No.421 Squadron at Chatham until November 1953, then with No.442 City of Vancouver Squadron until 1956. It was stored until 1964, then transferred to the RCAF's historic aircraft collection at Rockcliffe. The Museum also holds the RCAF's sole Vampire I, but its wings have been cut, and while it could be restored for display, this has not been done.

De Havilland Canada DHC-1 B2 *Chipmunk 2*

The design of the Chipmunk was started in the fall of 1945 at the suggestion of the English parent company. W.J. Jakimiuk, DHC's Chief Designer, a Polish engineer who had come to Canada early in the war, was largely responsible for its design. Manufacturing started in December 1945 and the prototype first flew at Downsview, Ontario, on May 22, 1946, only about five months later.

The type was immediately accepted as a good elementary trainer and 217 were made in Canada. The majority of the Canadian-built Chipmunks went into use with the RCAF and with the flying clubs for use in a DND-sponsored refresher training program for wartime pilots. In addition Canadian-built Chipmunks were supplied to the air forces of Egypt, Lebanon and Thailand and a number of civil machines were exported. The Chipmunk became the first aircraft of Canadian design to be made abroad under licence, with 1,000 built in England and 60 in

SPECIFICATIONS

Powerplant: One 145 hp
　　　　　D.H. Gipsy Major 10
Wing span: 34 ft 4 in (10.5 m)
Length: 25 ft 5 in (7.7 m)
Height: 7 ft (2.1 m)
Wing area: 172.5 sq ft (16 m²)
Weight, empty: 1,199 lb (544 kg)
Weight, gross: 1,930 lb (875 kg)
Speed, max.: 140 mph (225 km/h)
Speed, cruising: 124 mph
　　　　　(200 km/h)
Initial rate of climb: 900 ft/min
　　　　　(275 m/min)
Service ceiling: 17,200 ft (5240 m)

Portugal. British-built Chipmunks were supplied to the RAF and to air forces in several other sterling countries. The Duke of Edinburgh trained on the type, which gained it considerable publicity.

When the Chipmunk reached the end of its career as a military trainer, surplus machines found wide acceptance with private owners who wanted a nimble aircraft for personal use, and many remain in civil use today. Some were re-engined and otherwise modified to enhance the type's already excellent aerobatic capability and one of these appeared frequently at air shows in the hands of aerobatic champion the late Arthur Scholl.

The Museum's specimen, CF 12070, formerly RCAF 18070, the 207th made in Canada, was taken on strength in September 1956 and served at Centralia and Dunnville, Ontario; Portage, Manitoba; and Saskatoon, Saskatchewan, before being retired in 1971. It was purchased from Crown Assets in April 1972 and given the civil identity CF-CIA for the flight to Rockcliffe on April 25, 1972. It remains airworthy but has only once been used by the Museum for flight demonstration.

De Havilland Canada
DHC-2 Beaver 1

SPECIFICATIONS

Powerplant: One 450 hp Pratt &
        Whitney R-985
        AN-14B Wasp Jr.

Wing span: 48 ft (14.6 m)

Length: 30 ft 4 in (9.2 m)

Height: 9 ft (2.7 m)

Wing area: 250 sq ft (23.2 m²)

Weight, empty: 2,850 lb (1293 kg)

Weight, gross: 5,100 lb (2313 kg)

Speed, max.: 160 mph (258 km/h)

Speed, cruising: 130 mph (209 km/h)

Initial rate of climb: 1,020 ft/min
                  (311 m/min)

Service ceiling: 18,000 ft (5490 m)

The design study of a new bush or utility aircraft was started by de Havilland of Canada before the end of World War II but was put aside to complete the design of the DHC-1 Chipmunk. The idea was revived in 1946 and the final design was tailored closely to the requirements of the Ontario Provincial Air Service. Detail design began in September 1946 and part manufacturing started in December. The Beaver prototype was first flown on wheels at Downsview, Ontario, on August 16, 1947, by Russell Bannock.

The Beaver was the first all-metal Canadian bush aircraft. Its very effective flaps combined with drooped ailerons gave it a fine STOL performance, and its flying qualities, combined with its simple and rugged construction, were the major factors in its subsequent adoption for service in many countries. The Beaver became the most numerous of all Canadian-designed aircraft.

The most significant event in the Beaver's history was winning both the USAF and US Army competitions for a utility aircraft in 1951, which led to the delivery of 980 Beavers to the two services. Many served in Korea, where it was often known as the "general's jeep." The success of the Beaver marked a turning point in the history of DHC and the beginning of their STOL aircraft family. The Beaver was adopted by 13 other air forces as well as being put to civil use in many countries. As a result Beavers could be seen hard at work on all continents as well as in the Arctic and Antarctic regions.

The Beaver never entered Canadian military service but was widely adopted by bush operators from coast to coast, with the largest user being the Ontario Provincial Air Service. In later years, when the type was retired from military use, many examples were repatriated with the result that in recent years more Beavers were flying in Canada than ever before.

The Museum's specimen is, appropriately, the prototype, CF-FHB, which landed at Rockcliffe on the Ottawa River on July 29, 1980, after almost 33 years service in western Canada. After about a year's use with its maker, it became the sixth Beaver to be sold when Central B.C. Airways bought it in May 1948. 'FHB continued in their service and with their successors, Pacific Western Airways, until March 1966. It flew with Northward Aviation and B.&B. Aviation of Edmonton until April 1969, when it joined North Canada Air (Norcanair) of Prince Albert, Saskatchewan, remaining with that company until bought by the Museum in 1980.

De Havilland Canada DHC-3 Otter

SPECIFICATIONS

Powerplant: One 600 hp Pratt & Whitney S1H1-G Wasp

Wing span: 58 ft (17.7 m)

Length: 41 ft (12.5 m)

Height: 13 ft (4 m)

Wing area: 375 sq ft (34.8 m²)

Weight, empty: 5,287 lb (2398 kg)

Weight, gross: 8,000 lb (3628 kg)

Speed, max.: 160 mph (258 km/h)

Speed, cruising: 138 mph (222 km/h)

Initial rate of climb: 1,000 ft/min (305 m/min)

Service ceiling: 17,900 ft (5460 m)

Following the success of the Beaver, the Ontario Provincial Air Service decided they needed a larger machine with similar abilities and told de Havilland Canada that, if the company built an aircraft of about twice the Beaver's capacity and with similar performance, the OPAS would order 20. The new machine, originally named the King Beaver and later the Otter, was ordered into production in November 1950. It could carry up to nine passengers or 3,153 lb (1430 kg) of payload and had a

137

large double freight door and an air-drop hatch at the rear of the cabin. The prototype first flew on December 12, 1951, and was certified the following November.

The Otter, like the Beaver, had excellent STOL characteristics and, unlike the Beaver, had full-span flaps. Like the Beaver too, the Otter entered civil and military service in many countries. Military users included the air services of Argentina, Australia, Burma, Canada, Chile, Great Britain, Ghana, India, Indonesia, Norway and the United States. With the Otter the US Army developed new concepts in moving and supplying troops; altogether the US Army operated 161 Otters, designated YU-1 and U-1 in their service.

In Canada the Otter was widely used in bush flying, with the OPAS being the largest operator. The OPAS used Otters to develop water bombing techniques for fighting bush fires, with three different types of tank installation being used. Oversize low-pressure tires were used for landing on the barren Arctic Islands by Bradley Air Services, as had been done with the Beaver too. The RCAF Otters were largely used for search and rescue work and some were loaned to the United Nations Organization for use in operations in the Near East. In later years the RCAF and CF Otters were flown by the reserve squadrons.

The Museum's specimen was taken on strength by the RCAF in August 1960 and served with Nos. 403 and 408 (Aux.) squadrons and No.4 Search and Rescue Unit in Alberta until 1975. It was then transferred to No.1 Search and Rescue Unit at Montreal until the type was retired in 1982. It was refinished in RCAF Air Transport Command markings in 1983 by 6 RD, Trenton, and presented to the Museum in September 1983.

✳

Douglas DC-3

The venerable and apparently everlasting DC-3 is one of the world's great aircraft and the type that firmly established the airline transport industry as a profitable enterprise. Its story begins with the design of the DC-1 in 1933 to suit the requirements of Transcontinental and Western Air; the DC-1 entered production as the DC-2. The DC-2 was further enlarged and improved to enter American Airlines service in 1936 as the DST (Douglas Sleeper Transport) with 14 berths for passengers on transcontinental service. It was then realized that a day transport version of the DST was required, which resulted in the DC-3, a 21-passenger version that also entered service in 1936.

The DC-3 was immediately and enthusiastically accepted and 430 were delivered to American and European airlines by the entry of the

SPECIFICATIONS

Powerplant: Two 1,200 hp Pratt
      & Whitney R-1830
      Twin Wasp
Wing span: 95 ft 6 in (29.1 m)
Length: 64 ft 5 in (19.6 m)
Height: 16 ft 11 in (5.2 m)
Wing area: 987 sq ft (91.7 m²)
Weight, empty: 18,300 lb (8300 kg)
Weight, gross: 25,200 lb (11 430 kg)
Speed, max.: 237 mph at 8,800 ft
      (381 km/h at 2680 m)
Speed, cruising: 170 mph (274 km/h)
Rate of climb: 1,100 ft/min
      (335 m/min)
Service ceiling: 23,000 ft (7010 m)

United States into the war, with a further 149 on order. The DC-3 had all the improvements introduced in the Boeing 247 plus wing flaps and a greater capacity. It was the first transport aircraft that could operate at a profit without a subsidy and so introduced many new passengers to the benefits of air transport. Like the Boeing 247, the DC-3 had a retractable undercarriage; they were the first two transports to do so, and both allowed the wheels to protrude slightly from the nacelles as a measure of protection in the event of a forced landing. This proved unnecessary and the next generation of transports dropped the feature.

The DC-3 has long relinquished its role as a first-line air transport to larger and faster types, but 50 years after its introduction large numbers remain in service in various capacities in many countries, and seem likely to carry on for some time to come. In Canada alone, just under a hundred remain in use. In military use the type acquired numerous names and nicknames, Dakota, Skytrain, Skytrooper, "Gooney Bird", etc., which are sometimes applied to the civil DC-3 in the absence of a formal civil designation. In Canada the RCAF introduced the military version, the C-47, in 1943, while the DC-3 first entered Canadian civil use with Trans-Canada Air Lines in 1945.

The Museum's specimen was on civil order as a DC-3 prior to Pearl Harbor and was impressed into the USAAF as a C-49J-DO transport. It was sold to TCA on May 21, 1945, converted to civil use by Canadair Ltd. and handed over to TCA on September 21, 1945, as its first DC-3.

Sold to Goodyear Tire and Rubber Co. of Canada on October 4, 1948, it was refinished as a VIP transport and its Wright Cyclone engines were replaced by Pratt & Whitney Twin Wasps. It continued in Goodyear's service until December 19, 1983, when it was generously given to the Museum. As the first Canadian DC-3 and with 38 years of Canadian service, it is the most appropriate of all DC-3s to represent the type in the National Aviation Museum.

Douglas C-47B Dakota IV

SPECIFICATIONS

Powerplant: Two 1,200 hp Pratt & Whitney R-1830-90C Twin Wasp

Wing span: 95 ft 6 in (29.1 m)

Length: 63 ft 9 in (19.4 m)

Height: 17 ft (5.2 m)

Wing area: 987 sq ft (91.7 m²)

Weight, empty: 18,135 lb (8226 kg)

Weight, gross: 26,000 lb (11 794 kg)

Speed, max.: 224 mph at 10,000 ft (361 km/h at 3050 m)

Speed, cruising: 160 mph (258 km/h)

Rate of climb: 10,000 ft (3050 m) in 9.5 min

Service ceiling: 26,400 ft (8050 m)

It is surprising to note that, although it was only two years after the DC-2 entered civil use that it entered US Army Air Corps service as the C-32, it was five years after the DC-3 entered civil use that its military counterpart, the C-47, entered USAAC service. The C-47 was the most numerous of all DC-3 variants, and in all a total of 10,654 DC-3 variants were made in the United States as well as 414 in Japan and over 2,000 in the USSR.

While never receiving the headlines accorded combat aircraft, the C-47 was nevertheless one of the great tools that made the Allied victory possible. Its role was mainly the airlifting of troops and supplies wherever they were urgently needed, and frequently the C-47 was the only type available to do the job. During the invasion of Europe and in other attacks, it was the C-47 that delivered paratroopers and towed gliders loaded with troops to the designated areas, and in the postwar period the C-47 participated in the Berlin airlift of 1948. After the war many C-47s were converted for civil use, approximately as DC-3s, and many of these fly today on a variety of duties.

The RCAF received its first C-47 in March 1943 and the type has continued in its service and that of its successor to this day. During that time it has served as the equipment, in whole or in part, of the following squadrons: 164, 165, 168, 408, 12/412, 13/413, 14/414, 435, 436 and 448. These units supplied transport and other support to Canadian forces in Canada and Europe and, in the case of No.435 Squadron, in Burma during World War II and in Canada following the war.

The Museum's specimen was built at Douglas Aircraft Co.'s Oklahoma City plant in 1944 as US Army Air Force 44-75590 and immediately transferred to the RAF as KN451. It was a C-47B, which indicated that it had supercharged engines intended originally for flying the Hump route to supply China from India via Burma. Details of its RAF career are not known but it was obtained by the RCAF in August 1947. It served on transport duties with Eastern Air Command until transferred to Camp Borden as a trainer in November 1951. It was transferred to the RCAF's historical aircraft collection at Rockcliffe in September 1964. Unfortunately it has had to be stored outside and some deterioration has resulted.

The development of the North Star started late in 1944 and its airframe design may be considered as a combination of the best features of the Douglas C-54, DC-4 and DC-6 aircraft for its contemplated Canadian use. The Rolls-Royce Merlin engine was selected as its powerplant by Trans-Canada Air Lines. The prototype first flew on July 15, 1946, at Cartierville, Quebec, and was christened North Star on the 20th. The name was later adopted as the type name in Canada but British Overseas Airways Corporation selected Argonaut as the type name for their specimens.

The RCAF North Stars were unpressurized and bore the C-54GM designation, while the civil North Stars were pressurized, flew at a higher gross weight and were designated DC-4Ms. After early problems were overcome, the type gave dependable service but it is best remembered by passengers for its engine noise. TCA devised a new exhaust system in 1953-54, which alleviated but did not eliminate the

Douglas/Canadair
C-54GM North Star 1 ST

Powerplant: Four 1,760 hp
            Rolls-Royce
            Merlin 622

Wing span: 117 ft 6 in (35.8 m)

Length: 93 ft 9-1/2 in (28.6 m)

Height: 27 ft 6 in (8.4 m)

Wing area: 1,462 sq ft (136 m²)

Weight, empty: 43,500 lb (19 730 kg)

Weight, gross: 73,000 lb (33 113 kg)

Speed, max.: 353 mph at 23,900 ft
            (568 km/h at 7285 m)

Speed, cruising: 325 mph at 22,600 ft
            (523 km/h at 6890 m)

Initial rate of climb: 972 ft/min
            (296 m/min)

Service ceiling: 36,000 ft (10 970 m)

noise problem. A new version with radial engines, known as the C-5, was developed by Canadair and used as a VIP transport by the RCAF but it did not enter production.

TCA used the type on its trans-Atlantic and domestic routes until its retirement in 1961. BOAC used the type on its routes from London to Africa and the Far East until it was retired in 1960 after almost 11 years of service. The RCAF used the type for general transport duties, and North Stars served with 426 Squadron stationed initially at Dorval, Quebec, and later at Trenton, Ontario. In 1950-54, 426 Squadron assisted the USAF by airlifting supplies to Korea across the Pacific. North Stars also served with 412 Squadron at Uplands, Ontario, where they were mainly used as VIP transport for Canadian government officials and visiting dignitaries. The RCAF North Stars were retired in 1966.

The Museum's specimen was taken on strength by the RCAF in March 1948 and served with 426 Squadron until retired to the Museum in 1966.

Fairchild FC-2W-2

By the mid-1920s the Fairchild Aerial Survey Co. had become dissatisfied with the available aircraft types and formed the Fairchild Airplane Manufacturing Co. on Long Island, New York, to design and build an aircraft more suitable to their purposes. Their prototype, the FC-1,

SPECIFICATIONS

Powerplant: One 420 hp Pratt &
Whitney Wasp B
Wing span: 50 ft (15.2 m)
Length: 33 ft 2 in (10.1 m)
Height: 9 ft 6 in (2.9 m)
Wing area: 310 sq ft (28.8 m²)
Weight, empty: 2,732 lb (1239 kg)
Weight, gross: 5,500 lb (2495 kg)
Speed, max.: 134 mph (216 km/h)
Speed, cruising: 108 mph (174 km/h)
Initial climb: 875 ft/min (267 m/min)
Service ceiling: 15,000 ft (4570 m)

first flew in June 1926 and proved their ideas. The production model, the FC-2, was flown and marketed in 1927. It was well received, particularly in Canada, where Fairchild's Canadian company had a voice in its design, and the type was adopted by the RCAF and Canadian civil aircraft operators.

The RCAF shortly issued a requirement for an enlarged version and Fairchild improved the design by doubling the power and increasing the wing span. The new version was the FC-2W, which appeared in 1928 and with more minor improvements became the FC-2W-2, and with still more minor improvements the 71. In Canada it was these three very similar models that did so much, possibly the lion's share, of the outstanding work that opened up Canada's northern areas in the late 1920s and through the 1930s.

While the RCAF Fairchilds were used mainly in photographic survey operations, those in civil use were largely used for northern freighting operations in mineral exploration and developing mine sites. They were also used on air mail and passenger routes in southern Canada, and the first Canadian international passenger service, which connected Montreal and New York, used FC-2W-2s and 71s. Two Canadian FC-2Ws were much in the headlines in the spring of 1928 when they went to the rescue of the crew of the Junkers W.33 *Bremen,* who were marooned on Greenly Island in the Strait of Belle Isle after the first east-west Atlantic flight.

The Museum's specimen was built in mid-1928, registered as NC6621 and sold to Brock and Weymouth Inc. of Philadelphia, and continued in survey work until early in the 1940s when it was retired in favour of new equipment. It and its sister ships were sold to an aircraft broker, but on take-off NC6621 suffered an engine failure and was damaged in the ensuing forced landing. It remained in the company's possession until 1962, when Virgil Kauffman, founder and president of Aero Service, the oldest aircraft operating company in North America, generously gave it to the Museum. It has been restored in the markings of G-CART of Canadian Transcontinental Airways and was placed on display in 1966.

*

Fairchild 82A

SPECIFICATIONS

Powerplant: One 600 hp Pratt & Whitney Wasp R-1340 AN-1 (450 hp Pratt & Whitney Wasp SC-1)

Wing span: 51 ft (15.5 m)

Length: 36 ft 10-3/4 in (11.2 m)

Height: 9 ft 4-1/2 in (2.8 m)

Wing area: 343 sq ft (31.9 m²)

Weight, empty: 3,060 lb (1388 kg)

Weight, gross: 6,000 lb (2720 kg)

Speed, max.*: 145 mph (233 km/h)

Speed, cruising*: 128 mph (206 km/h)

Climb*: 9,000 ft (2740 m) in 10 min

Service ceiling*: 15,650 ft (4770 m)

* Indicates original equipment and performance.

The Fairchild 82 was designed by Fairchild Aircraft Ltd. at Longueuil, Quebec, as a successor to the American-designed Fairchild FC-2W-2 and 71 models that had served so well in northern Canada. The design followed the same general type of construction and even used the 71C wings. However, it was appreciably improved by its new fuselage with a much enlarged cabin giving better load distribution and fitted with detachable bench seats as well as larger doors to facilitate cargo loading. The type was offered with several different powerplants of increased power, which enhanced its performance over its predecessors.

The prototype was first flown from the St. Lawrence River at Longueuil on July 6, 1935, and it won immediate acceptance by the bush operating companies. The type was also sold abroad, with seven 82s being exported to Argentina, Mexico and Venezuela. It was welcomed by its users, some of whom preferred it to the Noorduyn Norseman, which first flew in November 1935 and was ordered in quantity during World War II. However, Fairchild phased out production of the 82 to make way for an RCAF contract to manufacture the Bristol Bolingbroke, with the result that only 24 were made. Fairchild hoped to revive 82 production in the immediate postwar period but these hopes were dashed when it was found that the 82 tooling had been scrapped during the war.

The Museum's specimen, an 82A, CF-AXL, the 18th built, was sold in March 1937 to Starratt Airways and Transportation Ltd. of Hudson, Ontario. It remained in their service until the company was sold to the Canadian Pacific Railway and became part of Canadian Pacific Airlines in May 1942. In 1947 'AXL was sold to Ontario Central Airlines of Kenora, Ontario, remaining with them until 1952. It then passed through the hands of a number of smaller operators in Alberta and Saskatchewan until acquired by CPA again in 1965. Sometime in the postwar period the original Pratt & Whitney SC-1 engine was replaced by the wartime R-1340 AN-1 Wasp, which developed more power. The aircraft was refinished in the wartime markings of CPA and handed over to the Museum on May 16, 1967, during a special CPA 25th Anniversary Dinner held in Ottawa. It is one of two surviving 82s, the other being an Argentinian machine held by the Musea Nacional de Aeronautia in Buenos Aires.

While the D.H.82C Tiger Moth and the Fleet 16 Finch served the RCAF and the British Commonwealth Training Plan well in the early days of World War II, both designs were obsolete even at the beginning of the war. A new aircraft was needed with flying characteristics more similar to service aircraft of the time and with more complete instrumentation to aid the student pilot in making the transition to more advanced types.

Fairchild PT-26B Cornell III

SPECIFICATIONS

Powerplant: One 200 hp
Ranger 6-440-65
Wing span: 36 ft 11-3/16 in (11.3 m)
Length: 28 ft 8 in (8.7 m)
Height: 7 ft 7-1/2 in (2.3 m)
Wing area: 200 sq ft (18.6 m²)
Weight, empty: 2,022 lb (917 kg)
Weight, gross: 2,736 lb (1241 kg)
Speed, max.: 122 mph (196 km/h)
Speed, cruising: 101 mph (162 km/h)
Initial climb: 645 ft/min (197 m/min)
Service ceiling: 13,200 ft (4020 m)

After a demonstration in Ottawa the RCAF decided on a new version of the Fairchild PT-19 being built in the United States for the USAAF. The main differences required for the Canadian machine were the addition of a cockpit enclosure and improved heating system. An order for the new type was given to Fleet Aircraft Ltd. at Fort Erie, Ontario, in December 1941. Altogether Fleet produced 1,642 examples of the type, the PT-26A Cornell II for the RAF and the PT-26B Cornell III for the RCAF. Fleet's first Cornell, a PT-26B for the RCAF, was first flown on July 9, 1942.

The RAF's Cornells were used in their training schools in India and South Africa, while those of the RCAF replaced the earlier trainers in the BCATP elementary schools as soon as they became available. Twenty Fleet-built Cornells were used by the Royal Norwegian Air Force to supplement their American-built PT-19s and Cornell Is in use at their Little Norway base in Muskoka, Ontario.

In the postwar period Cornells were quickly phased out by the RCAF and none remained by mid-1947. Many surplus Cornells were placed in civil use by flying clubs and private owners, and while they remained in use for a number of years, they have not survived in the same numbers as the Tiger Moths.

The Museum's specimen was built in July 1942, was issued to No.4 Training Command at Regina, Saskatchewan, and served at one or more or their elementary schools until July 1944. It was then stored in western Canada until it was transferred to the RCAF's historic aircraft collection at Rockcliffe in 1964.

Fairey Swordfish I.T

SPECIFICATIONS

Powerplant: One 730 hp Bristol Pegasus XXX

Wing span, upper: 45 ft 6 in (13.9 m)

Wing span, lower: 43 ft 9 in (13.3 m)

Length: 35 ft 8 in (10.9 m)

Height: 12 ft 4 in (3.8 m)

Wing area: 607 sq ft (56.4 m²)

Weight, empty: 4,700 lb (2132 kg)

Weight, gross: 7,510 lb (3406 kg)

Speed, max.: 139 mph at 4,750 ft (224 km/h at 1448 m)

Speed, cruising: 115 mph at 5,000 ft (185km/hat1524m)

Climb: 5,000 ft (1520 m) in 10 min

Service ceiling: 9,500 ft (2900 m) approx.

The Swordfish was one of the remarkable aircraft of World War II in that, while its design originated in 1930, it served operationally throughout the war and, it is said, destroyed a greater tonnage of enemy shipping than any other Allied type. This surprising performance can only be understood when it is realized that it normally operated in areas beyond the range of more modern enemy landplanes.

The design originated with a specification for a torpedo bomber issued in 1930. The resulting aircraft, the Fairey TSR I, first flew in 1933; a modified version, the TSR II, first flew in 1934 and was given the name Swordfish and ordered into production in 1935. It had one fixed gun forward and one flexibly mounted gun aft and carried an 18-inch torpedo or an equivalent bomb load.

Swordfishes operated from British aircraft carriers and also from shore bases. They carried out numerous courageous and successful torpedo attacks, the best known being one that crippled the Italian fleet at Taranto on November 11, 1940. By the end of 1941 the Swordfish was confined to an anti-submarine role as a spotter and hunter from escort carriers and merchant aircraft carriers. As far as is known, no

Canadians were involved on these British Fleet Air Arm operations with Swordfishes.

In Canada Swordfish operations began in 1943 at the No.1 Naval Air Gunnery School at Yarmouth, Nova Scotia, which was a part of the British Commonwealth Air Training Plan. A second group flew from the Royal Navy station at Dartmouth, Nova Scotia, designated HMS Seaborn. The RN equipment was taken over by the RCAF in May 1945 and in May 1946 the 46 Fleet Requirement Unit 743 (RCN) was formed equipped with Swordfishes.

The Museum's specimen was bought in 1965 by the Canadian War Museum from an eccentric farmer, Ernest Simmons of Tillsonburg, Ontario. Simmons had bought a fleet of North American Yales, Fairey Swordfishes and one or two Westland Lysanders together with other surplus material following World War II. They were stored on his farm and jealously guarded and he is believed to have sold only two in his lifetime. The balance of his collection was sold by his estate after his death. The specimen was restored by Fairey Aviation Co. of Canada Ltd. and the Royal Canadian Navy. As the aircraft's original identity was lost during its years at Tillsonburg, a serial number was arbitrarily assigned.

Fairey Battle IT

The Battle was designed to a 1932 specification for a day bomber and the prototype was first flown in March 1936. The standard service Battle had a crew of three – pilot, bomb aimer/observer and radio operator/gunner. It had a fixed gun for the pilot and a flexibly mounted gun for the rear gunner and could carry 1,000 lb (454 kg) of bombs. It went into production and began to be issued to RAF squadrons early in 1937. The Mk. numbers allocated to the Battle followed those of their Merlin engines, so there were Mks. I, II and III. Later a two-seat trainer with separate tandem cockpits was made, sometimes known as the Battle T, and still later came a Battle target tower conversion known as the T.T.

A rear gunner of a Battle scored the first RAF combat victory of World War II on September 20, 1939, but a few days later heavy Battle losses showed the type was not suitable for modern warfare. After the winter "phoney war" period, Battles were ordered to make daylight bombing missions on May 10, 11, 12 and 14, 1940. The results can only be compared to the charge of the Light Brigade in the Crimea. Much bravery was displayed, medals were won, little was achieved and very heavy losses were incurred – 100 per cent on the afternoon of May 12.

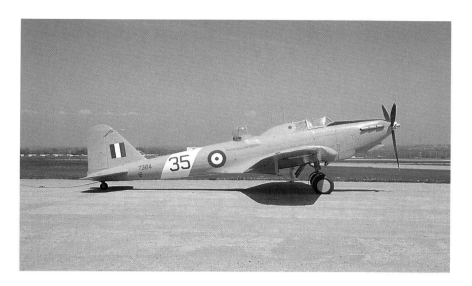

SPECIFICATIONS

Powerplant: One 1,030 hp
            Rolls-Royce Merlin III
Wing span: 54 ft (16.5 m)
Length: 42 ft 4 in (12.9 m)
Height: 15 ft 6 in (4.7 m)
Wing area: 422 sq ft (39.2 m²)
Weight, empty*: 6,647 lb (3015 kg)
Weight, gross*: 10,792 lb (4894 kg)
Speed, max.*: 210 mph (338 km/h)
           at sea level
Climb*: 5,000 ft (1520 m)
           in 4 min 6 sec
Service ceiling*: 25,000 ft (7620 m)

The type was then used briefly for night bombing. Three squadrons served for some time with Coastal Command, but the vast majority of Battles served out the war on training duties, with many being shipped from England to Australia and Canada.

In Canada 739 Battles served on training duties beginning in August 1939. A few trainer versions were used for pilot training but most served as target towers or gunnery trainers at all the bombing and gunnery schools of the British Commonwealth Air Training Plan. Following the war, Battles quickly disappeared from the RCAF's inventory, and were all gone by mid-1946. Very few appear to have been sold as surplus.

The Museum's specimen was taken on strength by the RCAF in January 1941 as a standard Battle I and served at No.31 Service Flying Training School, Kingston, Ontario, until December. It was then stored until converted to a gunnery trainer at St. Jean, Quebec, in December 1942 by fitting a Bristol Type I turret. It served briefly at No.3 Bombing and Gunnery School at Macdonald, Manitoba, and was stored in April 1943. It was repaired by the RCAF at Calgary in 1963 and transferred to the RCAF historic aircraft collection at Rockcliffe in 1964. It is the only Battle on display anywhere, although two museums have damaged or incomplete specimens intended for restoration.

* Figures for standard service Battle. The Museum's specimen, a gunnery training version, would vary only slightly.

Maurice Farman
Série 11 "Shorthorn"

SPECIFICATIONS

Powerplant: One 80 hp Renault
Wing span, upper: 15.78 m (51 ft 9 in)
Wing span, lower: 11.76 m (38 ft 7 in)
Length: 9.3 m (30 ft 6 in)
Height: 3.15 m (10 ft 4 in)
Wing area: 52 m² (559 sq ft)
Weight, empty: 654 kg (1,442 lb)
Weight, gross: 928 kg (2,046 lb)
Speed, max.: 116 km/h (72 mph)
Climb: 1000 m (3,280 ft) in 8 min

In the early days of aviation before World War I, pusher-type aircraft outnumbered the tractors, and a number of pushers played a prominent part in the Allied air forces in the early days of the war. These included designs of Henri and Maurice Farman, Gabriel Voisin, Geoffrey de Havilland and the Royal Aircraft Factory.

The Maurice Farman Série 11 was one such pusher design. The type was first flown in France late in 1913 and it was adopted by the air forces of Australia, Belgium, Britain, France, Italy and Russia and made under licence in Britain and Italy. Originally it was considered a standard service type and in the early period of the war it performed both reconnaissance and light bombing roles, without defensive armament for most of this time. It was then assigned to training duties in the service of the RFC/RAF until 1918 when it was finally retired from use. It was given the sobriquet Shorthorn to distinguish it from the Maurice Farman Série 7, which with its forward elevator was the Longhorn. The Série 11 was also known affectionately as the Rumpety for unrecorded reasons.

Only two Canadians are known to have flown the Shorthorn on operations, one in France and the other at Gallipoli. Canada had no air force at the time and most Canadian pilots became operational later in the war, after the Shorthorn's front-line career. However, many Canadians were trained on the type and still others instructed on it before its retirement. No Shorthorn is known to have entered Canada, or indeed North America, before the arrival of the Museum's specimen.

The Museum's aircraft was built by the Aircraft Manufacturing Co. (Airco) at Hendon, England, and was one of four shipped to Australia during World War I and used for flight training there at Point Cook. They were sold at auction in 1919 and two became registered civilly in Australia, with the others used for spares. The two aircraft were registered G-AUBC and G-AUCW, later becoming VH-UBC and VH-UCW, and were used mainly for joy riding. They were stored in the 1930s until the early 1950s, when 'UBC was rebuilt with some parts from 'UCW and flown. It was sold in 1956 to Frank Tallman of California, who flew it apparently only once or twice, and it was assigned US registration N9645Z. It was sold at auction in May 1968 to the Aeroflex Museum, which did not open, and shortly the specimen went to the Wings and Wheels Museum at Santee, South Carolina, and later at Orlando, Florida. It was sold at auction to the National Aviation Museum in December 1981. It still bears the Australian registration on the wings and it is intended to refinish it in RFC markings of World War I.

"Shorthorn" cockpit. Note the "spectacle" control column, believed unique, and the separate rudder pedals hinged at the heel with a cable at the toe. Instruments are, left to right, tachometer, airspeed indicator, oil pressure gauge, altimeter.
(KMM)

"Shorthorn" tail detail. Again note the many bracing and control wires, proving an old story that to check the rigging one placed a sparrow in it, and if it flew away there was a wire missing.
(KMM)

Farman S. 11 nacelle detail. The multitude of struts and wires and the height of the nacelle from the ground made entry into the cockpit difficult.
(KMM)

Fleet 16B Finch II

SPECIFICATIONS

Powerplant: One 125 hp Kinner B-5

Wing span,
upper and lower: 28 ft (8.5 m)

Length: 21 ft 8 in (6.6 m)

Height: 7 ft 9 in (2.2 m)

Wing area: 194.4 sq ft (18 m²)

Weight, empty: 1,122 lb (509 kg)

Weight, gross: 2,000 lb (907 kg)

Speed, max.: 104 mph (167 km/h)

Speed, cruising: 85 mph (137 km/h)

Climb: 1,000 ft (305 m) in 2.3 min

Service ceiling: 10,500 ft (3200 m)

The Fleet 16 was the final development of a light training aircraft designed by the Consolidated Aircraft Corp. of Buffalo, New York, in 1928 and intended for civil use. It was designated the Consolidated Model 14 Husky Jr., the name being an attempt to capitalize on the more powerful Consolidated PT-12 Husky. In fact, few if any were sold under the Consolidated or Husky names, as the type went into production under the Fleet name with various numerical model designations. Only the RCAF gave any of the models names, designating Models 7 and 16 as the Fawn and Finch respectively.

All the Fleet trainer models were very similar, with the powerplant usually being the main difference, but the Models 5, 10 and 16 had a fuselage deepened by fairings, new undercarriage and tail surfaces. The Fleet trainers were made in Canada from 1930 to 1941 by Fleet Aircraft of Canada at Fort Erie, Ontario. Models 2, 7, 10 and 16 were built there, and large numbers of the Model 10 were exported to nine countries.

The Model 16 was especially developed from the Model 10 to suit RCAF requirements. The airframe was strengthened locally and the only visible change was some additional wires. The Model 16 was made in two versions for the RCAF. The 160 hp Kinner-powered 16R was first flown in September 1939 and the 125 hp Kinner-powered 16B first flew in March 1940. A total of 437 Model 16s was built at Fort Erie, with the majority being supplied to the RCAF.

The 16B Finch II served in 12 of the elementary flying training schools during the early years of World War II and began to be replaced in 1943 by the Fairchild Cornell. Like their predecessors the Fawns, the

Finches were well liked, especially for their ruggedness. In 1940 a difficulty in recovering from an inverted spin became apparent but the trouble was diagnosed and corrected. Surplus 16Bs proved popular on the civil market for instruction and private flying and a few continue in use today.

The Museum's specimen was built in July 1940 and served with No.3 EFTS at London, Ontario, until August 1942, when it was transferred to No.11 EFTS at Cap-de-la-Madeleine, Quebec. Following its RCAF service, it was sold in the United States and registered civilly as NC1327V in New York. It was purchased late in 1963 by Dr. A.D. MacLean of Toronto and restored by 400 Squadron in its original RCAF markings. It was donated to the RCAF historic aircraft collection in June 1966 and was flown to Ottawa in September. It was flown at Rockcliffe in May and June 1967. A missing cockpit enclosure was donated in 1976 by Clifford Glenister of Toronto.

✳

Fleet 80 Canuck

SPECIFICATIONS

Powerplant: One 85 hp
Continental C-85-12J
Wing span: 34 ft (10.4 m)
Length: 22 ft 4-1/2 in (6.8 m)
Height: 7 ft 1 in (2.2 m)
Wing area: 173.5 sq ft (16.1 m²)
Weight, empty: 858 lb (389 kg)
Weight, gross: 1,480 lb (671 kg)
Speed, max.: 111 mph (179 km/h)
Speed, cruising: 100 mph (161 km/h)
Initial climb: 550 ft/min (168 m/min)
Service ceiling: 12,000 ft (3660 m)

The design of the Fleet 80 originated with J.O. (Bob) Noury of Noury Aircraft Ltd., Stoney Creek, Ontario. It was one of a pair of similar two-seat aircraft designed for civil use toward the end of World War II, one being a tandem-seated machine, the Noury T-65, and the other the side-by-side-seated Noury N-75. The N-75 was first flown at Hamilton late in 1944 and its design rights were sold to Fleet in May 1945.

Fleet began testing the machine in June and made a number of changes, the main ones being a metal wing structure, an enlarged vertical tail and an 85 hp engine. Marketing began early in 1946 and it was now named the Canuck after the sobriquet of the World War I Curtiss JN-4 (Can.). Fleet also designed floats for the type.

The Canuck proved popular with its rugged construction, generous baggage allowance, self starter, fuel-injection engine and other features. Its 1947 selling price of $4,869 plus $247 tax included the optional starter and green skylight. Flying clubs, fixed-base operators and private owners were the main purchasers in Canada, and 24 were exported: 19 to Argentina, three to Brazil and one each to Portugal and the United States. 225 were built before production stopped in 1949. Many continue in service today, with 93 listed in 1986 on the Canadian civil register. The Portuguese Canuck has migrated to England and presumably a few South American machines remain active.

The Museum's specimen was the 149th Canuck built. It was delivered in September 1947 to Canuck dealer Roger Watson, who sold it to Central Airways. That company was owned by Robert and Tommy Wong and operated from the Toronto Island Airport. It was the largest flight training operation in Canada and is believed to have operated more Canucks than any other Canadian operator. After almost 24 years service with Central Airways, the aircraft was sold in 1971 to Dr. J.D. Robinson of Flesherton, Ontario, who after two years passed it on to Ernest Weller of Port Loring, Ontario, from whom the Museum bought it in May 1974. The specimen is complete with wheels, Fleet floats and skis. The Museum plans to exhibit it on floats but it has been stored to date due to lack of space and it needs to be brought up to good display condition.

Fokker D.VII

The Fokker D.VII was the most widely publicized German fighter aircraft of World War I. After winning the fighter competition in January 1918, it was ordered into production both by Fokker and by Albatros Werke G.m.b.H. at the latter's Johannisthal (Alb.) and Schneidemühl (Ostdeutsche Albatros Werke) factories; more D.VIIs were built by Albatros than by Fokker. The D.VII was also built under licence in Austria.

On its introduction into service the type proved popular for its

SPECIFICATIONS

Powerplant: One 160 hp Daimler
Mercedes D.III av
Wing span, upper: 8.9 m (29 ft 2 in)
Wing span, lower: 7.0 m (23 ft)
Length: 7.0 m (23 ft)
Height: 2.75 m (9 ft)
Wing area: 21.5 m² (231 sq ft)
Weight, empty: 700 kg (1,544 lb)
Weight, gross: 850 kg (1,874 lb)
Speed, max.: 186 km/h (116 mph)
Climb: 1000 m (3,280 ft) in 3.8 min
Ceiling: 6980 m (22,900 ft)

constructional features and flying qualities. Its cantilever wooden wings and steel-tube fuselage proved easy to rig and repair and resistant to battle damage. It was easy and forgiving to fly and was particularly good at high altitudes, though not especially fast for the period.

It gained enduring fame by being singled out for mention in the Versailles Treaty and in the postwar period it was used by the air services of several countries, including Belgium, Canada, Finland, Holland, Latvia, Poland, Sweden and the United States. It was used by the short-lived Canadian Air Force in England in 1919 and within Canada it was only used briefly in 1919 when some of the war trophy aircraft were used in flying displays at the Canadian National Exhibition at Toronto, and one was flown in the New York-Toronto Air Race by Major W.G. Barker, VC.

The Museum's specimen was almost certainly one of the 142 brought to the United States and used briefly by its Air Service and later sold for civil use. It had a 200 hp Hall-Scott engine installed, and when civil aircraft registration was introduced in the United States in 1927, it became identified as 1178. It appeared in several Hollywood films, including the famed *Hell's Angels*, and was later bought by J. Mathiesen and J. Nissen, from whom it was purchased in February 1971 after the death of Mr. Mathiesen.

During restoration (see page 239), it was established that the

Museum's specimen was serial 10347/18 of the German Air Service, a very late production D.VII by Fokker probably taken over by the Allies at the factory before delivery. It is believed to be the only Fokker-built D.VII extant. Restoration was stopped in 1975 when well advanced to permit proceeding with the Curtiss HS-2L restoration and is expected to resume shortly.

Found FBA-2C

SPECIFICATIONS

Powerplant: One 290 hp Lycoming
    0-540-A1D
Wing span: 36 ft (11 m)
Length: 26 ft 5 in (8.1 m)
Height: 8 ft 4 in (2.5 m)
Wing area: 180 sq ft (16.7 m²)
Weight, empty: 1,550 lb (703 kg)
Weight, gross: 2,950 lb (1338 kg)
Speed, max.: 147 mph (237 km/h)
Speed, cruising: 129 mph (208 km/h)
Climb: 1,100 ft/min (335 m/min)
Service ceiling: 16,000 ft (4880 m)

About the end of World War II, two TCA captains, N.K. (Bud) and S.R. (Mickey) Found, started developing an idea they had held for some time, which was that a somewhat smaller bush or utility aircraft than usually available would fill a need of bush operators. A prototype, the FBA-1A, was built and flown in July 1949 by their company, Found Brothers Aviation Ltd.

The development of a production version was delayed several years due to financing and other problems but the prototype, the FBA-2A, was finally completed and flown at Malton, Ontario, on August 11, 1960. Two production versions were planned, the FBA-2B with tricycle undercarriage and the FBA-2C with a tail wheel. The FBA-2B was not proceeded with but the prototype FBA-2C first flew on May 9, 1962.

The FBA-2C was of all-metal construction and accommodated five people. Twenty-six FBA-2Cs were made and they went into service with various fixed-base operators across Canada, the largest operator being Georgian Bay Airways of Parry Sound, Ontario. In service the type neither distinguished itself nor disgraced itself and consequently some liked it and some did not. Ten FBA-2Cs were listed in the Canadian Civil Aircraft Register in 1986. The Museum's specimen was the third FBA-2C made and was originally registered CF-OZV in 1963. The final certification tests were carried out on it, and Found Brothers later used it as a demonstrator; it was never sold and placed in commercial service. Following the sale of most of the Found Brothers Aviation assets, the aircraft was bought by Centennial College, Scarborough, Ontario, in late 1969. The college gave it the fictitious registration C-GCCF and used it for ground instruction, including engine run-ups, until donating it to the Museum on April 10, 1979.

Hawker Hurricane XII

SPECIFICATIONS

Powerplant: One 1,300 hp
Rolls-Royce
(Packard) Merlin 29

Wing span: 40 ft (12.2 m)

Length: 32 ft 3 in (9.8 m)

Height: 13 ft 1-1/2 in (4 m)

Wing area: 257 sq ft (23.9 m²)

Weight, empty: 4,982 lb (2259 kg)

Weight, gross: 6,666 lb (3023 kg)

Speed, max.: 330 mph at 17,000 ft
(531 km/h at 5180 m)

Climb: 2,520 ft/min (768 m/min)

Service ceiling: 36,000 ft (10 970 m)

The Hurricane was built to a 1934 specification that was written around an earlier Hawker proposal, and the prototype was first flown in November 1936. Production began very quickly after the first order was given in June 1937, with the result that the first squadron began receiving machines in December 1937. And so it was that upon the commencement of the Battle of Britain there were 26 squadrons of

Hurricanes ready to meet the onslaught, although some were only partially equipped by that time.

It was the Hurricane that bore the brunt of the attack, as the famed Spitfire was only just entering squadron service. The Hurricane went on to be used on more fronts than any other British fighter – it was even used by the Russians – and in addition Sea Hurricanes carried out convoy protection duties by being catapulted from ships at sea, which usually entailed the pilots parachuting into the sea at the end of their missions. The type must go down in history as one of the most, possibly *the* most versatile fighter of World War II.

The RCAF received its first Hurricanes in August 1939, the air force's first modern fighters, and No.1 Squadron reached Britain just in time to take part in the Battle of Britain in Hurricanes. Commanded by S/L E.A. McNab, No.1 Squadron members were almost all from Montreal, and from the middle of August to mid-October 1940 they were credited with 31 enemy aircraft destroyed and 28 damaged for 16 Hurricanes lost, 3 pilots killed and 10 wounded or injured.

Britain in 1938 had given Canadian Car & Foundry Ltd. at Fort William, Ontario, a contract to make Hurricanes. The first Canadian-made Hurricane flew on January 10, 1940, and the plant went on to make 1,451 (including some Sea Hurricanes) before production stopped in the spring of 1943. Some Canadian machines arrived overseas in time to take part in the Battle of Britain and some were given to Russia. In Canada the RCAF used Hurricanes for defensive patrols on both coasts beginning in mid-1942 as well as for training.

The Museum's specimen, RCAF 5584, a 12-gun variant, was built by Canadian Car & Foundry at Fort William in November 1942 and was issued to No.2 Training Command at Winnipeg before going on to Western Air Command at Vancouver in March 1943, where apparently it had little use and was held in stored reserve. It was transferred to Portage la Prairie, Manitoba, in July 1950 and then on to 6 Repair Depot, Trenton, in May 1958 before finally being shipped to the RCAF's historic aircraft collection at Rockcliffe in February 1964. It is one of five known Hurricanes in Canada and one of six Canadian-built Hurricanes known to survive.

Hawker Sea Fury F.B.11

Fury and Sea Fury fighter development started in 1943. These designs were intended to be smaller versions of the Hawker Tempest then entering production. The Fury was to be developed by Hawker and the

SPECIFICATIONS

Powerplant: One 2,480 hp Bristol
Centaurus 18
Wing span: 38 ft 4-3/4 in (11.7 m)
Length: 34 ft 8 in (10.6 m)
Height: 16 ft 1 in (4.9 m)
Wing area: 280 sq ft (26 m²)
Weight, empty: 9,240 lb (4190 kg)
Weight, gross: 12,500 lb (5670 kg)
Speed, max.: 460 mph at 18,000 ft
(740 km/h at 5490 m)
Climb: 30,000 ft (9140 m)
in 10 min 48 sec
Service ceiling: 35,800 ft (10 910 m)

Sea Fury by Boulton-Paul Aircraft. With the war about to end, the Fury contract was cancelled in February 1945 but Hawker later did make some for export. The Sea Fury contract was moved to Hawker, which flew the prototype in February 1945. Earlier production Sea Furies were Mk.X fighters and later production machines were Mk.11 F.B. fighter bombers able to carry 1,000-lb (454-kg) bombs or rockets under the wing. Of the 860 Sea Furies made, 725 production machines went to the Royal Navy, including 60 Mk.20s, a two-seat trainer version. The RN transferred some of their Sea Furies to the Royal Australian and Canadian navies. British and Australian Sea Furies served in Korea in the early 1950s and successfully engaged the Russian MiG-15 jet fighters.

The RCN acquired 74 Sea Fury F.B.11s from the RN beginning in 1948 and extending through to 1953. They served on HMCS *Magnificent* and on shore bases with the following squadrons: Nos. 803, 825, 871, 883, VX-10 and VT-40. No Canadian Sea Furies served in Korea. In May 1953 approval was given for the loan of 12 Canadian Sea Furies and 14 pilots to the British Fleet Air Arm for service in Korea, but the war ended before their training was completed. The Sea Furies gave good service with the RCN until they were retired in 1956 as the last piston-engined fighters to serve in the navy. Some Sea Furies still fly in modified form as civil racing aircraft.

The Museum's specimen was taken on strength by the RCN in May 1948. It served for most of its career with 871 Squadron, earlier designated 883 Squadron, and was based at HMCS *Shearwater* at Dartmouth, Nova Scotia, on the carrier HMCS *Magnificent* and, briefly, at Rivers, Manitoba. The specimen was stored for brief periods and it was twice involved in accidents necessitating repairs. It was sold in June 1956 to Bancroft Industries Ltd. of Montreal and stored at Fredericton, New Brunswick, until 1963, when it was given to the Canadian War Museum by the company. It was then restored by the RCN and Fairey Aviation Co. of Canada Ltd. at Eastern Passage, Nova Scotia, before being shipped to Rockcliffe in 1964.

✳

Heinkel He 162A-1 Volksjäger

SPECIFICATIONS

Powerplant: One 800 kg (1,764 lb) st BMW 003E-1

Wing span: 7.2 m (23 ft 7-1/2 in)

Length: 9 m (29 ft 6-1/4 in)

Height: 2.6 m (8 ft 6-3/8 in)

Wing area: 11.2 m² (120.6 sq ft)

Weight, empty: 1663 kg (3,667 lb)

Weight, gross: 2490 kg (5,490 lb)

Speed, max.: 835 km/h at 6000 m (519 mph at 19,690 ft)

Climb: 1280 m/min (4,200 ft/min)

Ceiling: 12 020 m (39,440 ft)

The remarkable Volksjäger or "People's Fighter" was conceived, designed and built in 90 days as a last-ditch effort to stave off Germany's defeat. It was proposed in the fall of 1944 as an inexpensive fighter suitable for construction by largely semi-skilled workers from non-strategic materials. Heinkel's proposal was accepted on September 24 over those from six other manufacturers and work started in the shops immediately, with detail drawings being made simultaneously, and all drawings were completed by October 29, a day ahead of schedule. Three assembly plants were designated to turn out the aircraft from components supplied by a number of makers and the total scheduled output was to be 4,000 machines per month. The prototype was first flown on December 6, 1944, at Schwechat. Its armament was two 30mm cannons.

It was planned to recruit young Germans and give them glider training, then convert them directly to the Volksjäger; to such dreaming Germany had apparently been reduced by the pressure of the Allied attack. Actually, the Volksjäger was no machine for the inexperienced pilot, as it proved unstable longitudinally and needed the full attention of an experienced pilot. The Luftwaffe accepted a number of Volksjägers in the early part of 1945 but there is no evidence that the type was ever used operationally. So while the aircraft is an interesting example of what can be done in an extreme emergency and was one of the first jet fighters, it had no influence on the outcome of the war.

The Museum holds two specimens of the He 162, the 75th and 86th production aircraft made at the Heinkel-Nord plant at Marienehe probably in late February or March 1945. Details of their German history are not known, although it is thought they had not gone to the Luftwaffe, nor is it known how they came to Canada. They were acquired by the Canadian War Museum from the RCAF. One specimen was repainted by the Canadian War Museum in 1966 in the markings of an aircraft of 3 JG/77 and the other remains unrestored in storage.

Hispano HA-1112-M1L Bouchón

The Messerschmitt Bf 109 was one of the great fighter aircraft of World War II and, like the Supermarine Spitfire, served from the beginning of the war to the end. Like the Spitfire too, it was developed through a number of versions. To say which was the better fighter is difficult to do briefly as it depends on what versions you are comparing and under what conditions. Both were very good; otherwise they would not have remained in production.

The Bf 109 prototype first flew in July 1935, powered with a Rolls-Royce Kestrel engine. The type then went into production as the Bf 109B with a Junkers Jumo 12-cylinder engine of 680 hp. It was this version that first saw combat in 1937, serving with the Spanish Nationalist air arm, and Switzerland adopted the type in 1938. The Bf 109E was the latest version when World War II began and was used in the Battle of Britain. The Luftwaffe lost 610 Bf 109Es in this battle, along with 235 twin-engined Bf 110s, while Britain lost 1,172 fighters (Hurricanes, Spitfires, Blenheims and Defiants). Later versions of the Bf 109 appeared and these were used on all fronts until hostilities ceased.

Powerplant: One 1,600 hp
Rolls-Royce
Merlin 500

Wing span: 9.93 m (32 ft 7 in)

Length: 9.04 m (29 ft 8 in)

Height: 2.59 m (8 ft 6 in)

Wing area: 16.2 m² (174.38 sq ft)

Weight, gross: 2859 kg (6,300 lb)

Weight, empty: 2445 kg (5,390 lb)

Speed, max.: 666 km/h at 3720 m
(414 mph at 12,200 ft)

Speed, cruising: 605 km/h (376 mph)

Climb: 1370 m/min (4,500 ft/min)
approx.

Service ceiling: 10 200 m (33,500 ft)

Spain continued to use this aircraft after trying its initial batch of Bf 109Bs and ordered small numbers of Bf 109Es, 109Fs and 109Gs from Germany. When the German supply dried up towards the end of the war, Spain obtained the manufacturing rights to the type. The machines were built by La Hispano Aviación S.A. at Seville and corresponded to the Bf 109G2 with different powerplants. The initial versions were powered with Hispano-Suiza 12-cylinder engines and later versions with Rolls-Royce Merlins. Both these engines were upright Vees, in contrast to the inverted Vee of the earlier Daimler-Benz, and this inevitably gave the Spanish machines a different appearance, especially the Merlin-equipped aircraft, which were named Bouchón or "Pouting Pigeon."

The first Spanish-built Bf 109, an HA-1109-J1L, first flew on March 2, 1945. Altogether nine different variants of the type were made in Spain, using the two powerplants, and they incorporated various installations of armament and equipment. Production ceased in the spring of 1961.

The Museum's specimen was aircraft 39 of 471 Squadron and was obtained by the makers of the British documentary film *Battle of Britain* but was not actually used. The Canadian War Museum purchased it in England in 1967. It needed considerable restoration and was not placed on display before this was started in 1984. The work was finished in early November 1986 and the aircraft is finished in its original Spanish markings.

Junkers J.I

SPECIFICATIONS

Powerplant: One 200 hp Benz Bz IV
Wing span, upper: 16 m (52 ft 6 in)
Wing span, lower: 10.85 m (35 ft 7 in)
Length: 9.1 m (29 ft 10-1/4 in)
Height: 3.4 m (11 ft 1-7/8 in)
Wing area: 49.4 m² (531.7 sq ft)
Weight, empty: 1766 kg (3,893 lb)
Weight, gross: 2140 kg (4,718 lb)
Speed, max.: 155 km/h (96 mph)
Climb: 2000 m (6,560 ft) in 32 min

Dr. Hugo Junkers had been interested in metal aircraft construction since 1910 when he patented his ideas. His first design was a mid-wing monoplane made of sheet iron and flown as a prototype only in 1915. His second design was a low-wing monoplane first flown in 1916, of which five were made.

His first aircraft to go into general use was designated J.I by the German Air Service and J 4 by Junkers. It was a large cantilever biplane of aluminum construction intended for contact patrol work with the troops, involving low flying along the front lines under intensive small arms fire. To provide protection Junkers encased the crew of two, the engine and fuel tanks with 5-mm-(1/5-in-) thick armour plate, and the typical Junkers wing with its multi-spar construction covered with corrugated aluminum sheet proved also to be highly resistant to ground fire. The armament of the Museum's specimen is one flexibly mounted gun for the observer, but other accounts state that fixed forward-firing guns were also provided.

The prototype was first flown in January 1917 and shortly entered service at the front, and by the Armistice 227 had been made. When it first entered service, crews tended to look down on its long take-off run and difficult landing characteristics, but they soon began to appreciate its sturdy construction and almost complete invulnerability to ground fire, and these same characteristics made it a respected adversary by the opposing troops.

The Museum's specimen was brought to Canada as a war trophy and like the A.E.G. G.IV it was shipped from Dieppe on the SS *Venusia* on May 28, 1919. At Toronto it was assembled and exhibited to the public at the Canadian National Exhibition in August (see page 269) and then shipped to Camp Borden. Also like the A.E.G. G.IV, it found its way to Ottawa and was offered to the Aeronautical Museum in 1939 and reported as "being in the Archives." It too came to be stored in the Canadian War Museum's warehouse and apparently came on their inventory "by right of possession." It was first displayed at Rockcliffe in 1964, still unrestored. While it will likely be restored, because of the sizable job of reskinning the wings this will probably be well in the future.

The Museum's specimen is the only complete Junkers J.I extant but a J.I fuselage exists in Italy.

Junkers W.34f/fi

SPECIFICATIONS

Powerplant: One 600 hp Pratt & Whitney Wasp R-1340 AN-1
Wing span: 17.76 m (58 ft 3-1/4 in)
Length: 10.27 m (33 ft 8-3/8 in)
Height: 3.19 m (10 ft 6 in)
Wing area: 44 m² (473.6 sq ft)
Weight, empty: 1655 kg (3,649 lb)
Weight, gross: 3200 kg (7,055 lb)
Speed, max.: 264 km/h (164 mph)
Climb: 2000 m (6,560 ft) in 9 min
Service ceiling: 6900 m (22,640 ft)

The Junkers W.34 was a direct development or modification of the Junkers F.13, a civil aircraft of which the prototype first flew in May 1919. The F.13 was hailed as the best civil aircraft of the immediate postwar period and went into service in many countries including Canada. It featured the all-metal construction developed by Junkers during the war, including the multi-spar wing and corrugated aluminum covering. All-metal construction was, of course, quite unusual at that time. In 1926 Junkers introduced two developments of the F.13, the W.33 and W.34, which both had a larger capacity fuselage. The W.33 was primarily a freighter and was usually sold with a Junkers L.5 or L.5G water-cooled engine, while the W.34 was primarily a passenger or passenger/freight machine and powered with a variety of air-cooled radial engines.

The two prototypes first flew in 1926 and, like the F.13, became widely accepted. The W.33 became world famous in 1928 when one, the *Bremen*, made the first east-west Atlantic crossing and landed on Greenly Island near the Quebec-Labrador boundary.

In Canada, Canadian Airways, with head office in Montreal and later Winnipeg, was the largest operator of the type, which it used for bush flying in eastern and western Canada. Canadian Airways stated it was the best bushplane of the period partly due to its good flying characteristics but mainly due to its rugged construction, which greatly reduced the cost of maintenance and repair. The fact that only nine W.33/W.34 machines were imported can be attributed partly to the increased cost of their metal construction and, probably mainly, to the heavy duty imposed on German aircraft.

The Museum's specimen was registered to Canadian Airways in December 1932. Its early years were spent based at Oskélanéo, Quebec, from where it flew north mainly to the Chibougamau area, servicing prospectors and helping to establish mines. In the summer of 1939 it was sent west to Lac du Bonnet, Manitoba, where it worked largely in the Red Lake area. Canadian Airways was sold to the CPR and became Canadian Pacific Airlines in May 1942 and 'TF continued at the same base until at least 1942, and it may well have remained there until 1946 when it was sold to Central B.C. Airways of Prince George, which became Pacific Western Airways in 1957. It was sold to Pacific Wings at Vancouver in 1960, then sold in 1962 to Mrs. J.A. Richardson, widow of James Richardson, founder of Canadian Airways, and donated by her to the Museum on September 17, 1962. This occasion marked the end of the active life of an early Canadian bushplane and also the last flight of any of the remarkable Junkers F.13/W.33/W.34 family.

Lockheed 10A Electra

The Lockheed 10 was the first aircraft to be designed by the Lockheed company following its reorganization in 1932 and also their first all-metal design. It was intended to follow the trend set by Boeing with its 247 and Douglas with its DC-2 but to suit airlines that needed a smaller aircraft. It accommodated a crew of two and ten passengers and had all the modern features then just appearing – retractable

SPECIFICATIONS

Powerplant: Two 450 hp Pratt &
              Whitney Wasp Jr.
              R-985

Wing span: 55 ft (16.8 m)

Length: 38 ft 7 in (11.8 m)

Height: 10 ft 1 in (3.1 m)

Wing area: 458.5 sq ft (42.6 m²)

Weight, empty: 6,454 lb (2927 kg)

Weight, gross: 10,300 lb (4672 kg)

Speed, max.: 202 mph at 5,000 ft
              (325 km/h at 1520 m)

Speed, cruising: 190 mph (306 km/h)

Climb: 1,140 ft/min (347 m/min)

Service ceiling: 19,400 ft (5910 m)

undercarriage, wing flaps, variable-pitch propellers – which combined to give it good single-engine performance. The prototype first flew on February 22, 1934, and the type is usually referred to by number rather than its name.

The Lockheed 10 was adopted by a number of US airlines, including Chicago and Southern, Continental, Delta, National, Northwest and Pan American. It was also exported to many countries. Amelia Earhart, the noted aviatrix, was flying a 10E when she was lost over the Pacific in July 1937, and a special variant, the XC-35, was used by the USAAC to conduct pioneering cabin pressurization experiments. The type served well and a number remain in use today.

The Lockheed 10A became the first modern airliner to serve in Canada in 1936 when Canadian Airways bought two for their Seattle-Vancouver route. They were also to serve as pilot trainers in anticipation of their operating the proposed trans-Canada airway, but instead Trans-Canada Air Lines was set up by the government. TCA bought Canadian Airways' machines, started operating with them and ordered three more. TCA made little use of their 10As as they were quickly replaced by the powerful Lockheed 14s, and the TCA 10As were taken over by the RCAF together with others. An interesting operation by a 10E carried out from northern Canada in 1937-38 was the search for the missing Russian flyers, Levanevsky and crew, which was conducted by Sir Hubert Wilkins with pilots H. Hollick-Kenyon and S.A. Cheesman.

The Museum's specimen was the first of the three 10As ordered by

TCA direct from Lockheed and their third 10A after the two Canadian Airways machines. It became RCAF 1526 and served from October 1939 until 1949. It then passed to a number of owners in the United States until bought by L. Koepke. It was piloted by Ann Pellegreno on a flight around the world, July 7-11, 1967, commemorating Amelia Earhart's ill-fated flight of 1937. In March 1968, it was bought by Air Canada, re-registered and refurbished as CF-TCA and presented to the Museum on October 14, 1968.

Lockheed 1329 Jetstar 6

SPECIFICATIONS

Powerplant: Four 3,000 lb (1,360 kg) st Pratt & Whitney JT12-6

Wing span: 54 ft 5 in (16.6 m)

Length: 60 ft 5 in (18.4 m)

Height: 20 ft 5 in (6.2 m)

Wing area: 542 sq ft (50.4 m²)

Weight, empty: 22,074 lb (10 013 kg)

Weight, gross (take-off): 40,921 lb (18 562 kg)

Speed, max.: 490 knots at 22,000 ft (907 km/h at 6710 m)

Climb: 3,300 ft/min (1005 m/min)

Service ceiling: 33,000 ft (10 060 m)

In 1956 the USAF made it known that it would purchase a proven light jet transport that met its requirements, providing funds were available. Lockheed started design work in January 1957 and their prototype first flew on September 4, 1957, only 241 days later. It was powered by two British Bristol Orpheus turbojets, an engine then intended for American production.

By 1958 the USAF budget had been cut and the plan to produce the Orpheus turbojet had been dropped. Nevertheless Lockheed decided to commit the type to production as a business jet in their Marietta, Georgia, plant. To suit the type for future USAF orders, the engines had to be American, and since there was no equivalent American engine to the Orpheus, Lockheed had to replace the two Orpheus powerplants in the prototype with four Pratt & Whitney JT12-6s. The first production aircraft, the Jetstar 6, first flew in the summer of 1960 and received its type approval a year later. Later other versions were built, some of which were supplied to the USAF. Altogether 204 Jetstars were made and about 90 per cent of them are still in service in various countries. While the Jetstar was an early business jet, it was not the first. That distinction went to France's Morane-Saulnier MS.760 Paris, which first flew on July 29, 1952.

Eight Jetstars have served in Canada. The first was bought by the T. Eaton Company and two were bought by the Department of Transport, which later acquired the Eaton Jetstar also. The DOT (later MOT) Jetstars served from 1962 to 1986, almost a quarter of a century.

In DOT service the Jetstar carried a flight crew of two, up to nine passengers and, when serving as a VIP transport, a flight attendant as well. The Jetstar was used for airway inspection but is best remembered as a VIP transport, carrying senior Canadian government officials and visiting dignitaries including members of the Royal family. The Museum's specimen was the first delivered to the DOT and was retired to the Museum on June 9, 1986.

Lockheed 12A Electra Jr.

SPECIFICATIONS

Powerplant: Two 450 hp Pratt &
Whitney Wasp Jr.
R-985
Wing span: 49 ft 6 in (15.1 m)
Length: 36 ft 4 in (11.1 m)
Height: 9 ft 9 in (3 m)
Wing area: 352 sq ft (32.7 m²)
Weight, empty: 5,765 lb (2615 kg)
Weight, gross: 8,400 lb (3810 kg)
Speed, max.: 225 mph (362 km/h)
Speed, cruising: 213 mph (343 km/h)
Climb: 1,400 ft/min (427 m/min)
Service ceiling: 22,900 ft (6980 m)

With the Lockheed 10 still on the drawing boards, it was felt that a still smaller version should be designed, using the same engines and consequently giving enhanced performance. This design was aimed at feeder line operators and private operators, including business concerns. It had all the same features as the model 10 but housed only two pilots and six passengers. As a result it was almost identical in appearance to the model 10 except for size and window arrangement. The prototype first flew on June 27, 1936.

The type did not become widely used by feeder line operators though some were sold to them in the United States and abroad, but it was well received by private operators, and a military version, the model 212, was sold to the US Army Air Corps and foreign air forces. Production of the type was terminated by war orders to Lockheed early in World War II. Probably the most interesting careers of any 12s were those of three supplied to Britain. Fitted with hidden cameras, they were used by Sidney Cotton on behalf of the British government to surreptitiously photograph portions of Germany before the outbreak of the war.

The Museum's specimen was the only model 12 to be registered in Canada before World War II. However, the RCAF imported ten used model 12s and one model 212 from the United States beginning in 1940. Following the war, several continued to fly in Canada with civil operators.

The Museum's specimen, CF-CCT, was bought by the newly established Department of Transport in May 1936 to provide it with a modern high-performance aircraft to survey the routes of the proposed Trans-Canada airway and to set up and calibrate its radio facilities. While this was its main job in its early years, it is best remembered for its Montreal-Vancouver flight on July 31, 1937. The flight was arranged to take the Minister of Transport, the Honourable C.D. Howe, and other government officials over the proposed airway route. The flight was widely publicized, usually as a dawn-to-dusk flight, which was an exaggeration as it left Montreal at 1:18 a.m. Piloted by J.H. Tudhope and J.H. Hunter and with five stops, it arrived at Vancouver at 6:29 p.m. CF-CCT served all its working life with the DOT. When it was heard that it was due for retirement in 1963 after 26 years, the Museum suggested it be added to the collection both as a fine example of a light transport of the period and as an aircraft that had contributed much to setting up Canada's airway system. CF-CCT was handed over to the Museum on September 23, 1963.

Jet aircraft entered military operational use in the closing months of World War II and in the postwar period they rapidly gained acceptance in most air forces. Initially experienced pilots converted directly to jet fighters but it was soon realized that a two-seat trainer was needed to facilitate the conversion, especially with inexperienced pilots. Lockheed developed a two-seat stretched version of their successful

Lockheed T-33AN Silver Star 3

Powerplant: One 5,100 lb
(2313 kg) st
Rolls-Royce Nene 10

Wing span without tip tanks:
37 ft 7-1/2 in (11.5 m)

Wing span with tip tanks:
42 ft 7 in (13 m)

Length: 37 ft 8-1/2 in (11.5 m)

Height: 11 ft 8 in (3.6 m)

Wing area: 238 sq ft (22.1 m²)

Weight, empty: 8,440 lb (3828 kg)

Weight, gross: 16,800 lb (7620 kg)

Speed, max.: 570 mph (917 km/h)

Speed, economical cruising: 190 mph
(306 km/h)

Climb: 20,000 ft (6095 m) in 8 min

Service ceiling: 47,000 ft (14 330 m)

P-80 Shooting Star fighter for the purpose. The prototype was initially known as the TP-80C and first flew in March 1948, and the first production aircraft, now called the T-33, was delivered the following August. The type was used by the USAF, USN and USMC and by many other smaller air forces and it continues in service today in some countries.

In the early 1950s the RCAF selected the T-33 as a jet trainer to meet their NATO commitment to train not only their own pilots but also those from other NATO countries. The initial order was for 20 Lockheed-built T-33As, which the RCAF designated the Silver Star 1. The RCAF also had Lockheed make a version with the more powerful Rolls-Royce Nene engine, which was delivered in November 1951 and designated the Silver Star 2. The Nene-powered variant was put into production by Canadair Ltd. at Montreal and designated the Silver Star 3. The prototype first flew on December 22, 1952, and by the time production stopped in March 1959, 656 had been supplied.

The name Silver Star never became popular and the type is known everywhere as the "T-bird." The T-bird has served the RCAF well as a pilot and armament trainer and is well liked by its pilots. Many Canadian T-birds were sent on to Bolivia, France, Greece, Portugal and Turkey as part of the mutual aid program. Probably the T-bird is best remembered by the Canadian public not as a fine trainer but as the mount of the Red Knight, the soloist of the Canadian aerobatic team, which appeared at air shows at Canadian centres from coast to coast.

The Museum's specimen was built by Canadair in January 1957. It served at No.1 Advanced Flying School, Saskatoon, and, painted brilliant red, appeared as the Red Knight. The aircraft was retired to the Museum in May 1964, and it seems fitting that the Red Knight was chosen to represent the type.

Lockheed F-104A Starfighter

SPECIFICATIONS

Powerplant: One General Electric
J-79-GE-3B, 9,600 lb
(4354 kg) st thrust /
14,800 lb (6712 kg) st
thrust with afterburner

Wing span: 21 ft 11 in (6.7 m)

Length: 54 ft 8 in (16.7 m)

Height: 13 ft 6 in (4.1 m)

Wing area: 196.1 sq ft (18.2 m²)

Weight, empty: 13,384 lb (6070 kg)

Weight, gross: 17,988 lb (8159 kg)

Speed, max.: 1,037 mph (1669 km/h)

Speed, cruising: 519 mph (835 km/h)

Climb: 60,395 ft/min (18 408 m/min)

Service ceiling: 64,800 ft (19 750 m)

The Lockheed Aircraft Corp. started the design of the F-104 in an attempt to halt the trend to heavier and heavier fighters that had begun in the post-World War II years. The prototype first flew on March 4, 1954, at Edwards Air Force Base, California. It went into production as the slightly modified F-104A, the prototype of which first flew on February 17, 1957. The type proved highly successful and was built in seven countries – Belgium, Canada, West Germany, Italy, Japan, the Netherlands and the United States. In view of the fact it was intended as a lightweight fighter, it is ironic that its most successful version was the F-104G, originally made for Germany with all-weather radar, external fuel, provision for external stores, etc., which increased its weight considerably.

The excellent performance of the F-104 allowed it to set a number of new records, including a new world's altitude record of 103,389 ft (31 513 m), several speed records over prescribed courses and new time-to-climb records. In May and June 1964 Jacqueline Cochran flew a F-104 to better her own three women's speed records set in 1952 in the F-86 Sabre 3.

The RCAF selected the F-104 to replace the F-86 Sabre in West Germany and a manufacturing contract was given to Canadair Ltd. at Montreal. The Canadair-built machines were designated as CF-104s. The engine was also built in Canada, by Orenda Engines Ltd., and was designated as the General Electric J-79-OEL-7. The aircraft carried some RCAF-specified equipment, including a ventral reconnaissance pod with four Vinten cameras and special electronic sensors.

As the Starfighter was the RCAF's NATO fighter, only one squadron in Canada, No.448, the test unit at Cold Lake, Alberta, was equipped with it. Overseas, eight squadrons, Nos. 421, 422, 427, 430, 434, 439, 441 and 444, of 1 Air Division used the Starfighter. It was well liked in spite of newspaper accounts implying that it was crash prone. It seems that such crashes that have occurred can be attributed to the type of flying being done, weather conditions and, possibly, training.

The Museum's specimen, RCAF 12700, later 104700, was an F-104A, USAF 56-770, which came to Canada as a pattern aircraft for the Canadair examples that followed. It was flown at the Central Experimental and Proving Establishment at Uplands on test, and from there on December 14, 1967, W/C R.A. White flew it to a new Canadian altitude record of 100,110 ft (30 500 m). As it was an orphan and not suitable for squadron use, it was given to the Museum in June 1968.

McDonnell F2H-3 Banshee

SPECIFICATIONS

Powerplant: Two 3,200 lb (1450 kg) st Westinghouse J34-WE0-34

Wing span: 41 ft 9 in (12.7 m)

Length: 48 ft 2 in (14.7 m)

Height: 14 ft 6 in (4.4 m)

Wing area: 294 sq ft (27.3 m²)

Weight, empty: 13,183 lb (5980 kg)

Weight, gross: 21,013 lb (9531 kg)

Speed, max.: 580 mph (933 km/h)

Speed, cruising: 461 mph (742 km/h)

Climb: 6,000 ft/min (1829 m/min)

Service ceiling: 46,600 ft (14 200 m)

The Banshee was ordered by the US Navy in March 1945 and the prototype first flew in January 1947. It was basically an enlarged and more powerful version of the McDonnell FH-1 Phantom, which first flew in January 1945 and was the USN's first jet fighter. The Banshee was developed through nine versions and first entered USN service in 1948. 895 Banshees were made and the type saw active service in Korea as a bomber escort fighter. The Banshee was twin-engined and armed with four 20mm cannons and the F2H-3 and -4 versions were capable of operating in the all-weather role.

In Canada the Banshee was selected as the replacement for the navy's Hawker Sea Fury fighter, which was being retired. Thirty-nine Banshees were acquired from the USN, the first in November 1955 and the last in June 1958. The RCN Banshees served with fighter squadrons VF-870 and VF-871 and with experimental unit VX-10. They were entirely land-based at HMCS *Shearwater* at Dartmouth, Nova Scotia, until the carrier HMCS *Bonaventure* was commissioned in 1957. The Banshee was the RCN's first jet fighter and its last fighter, as it was not replaced upon its retirement in the fall of 1962.

The Museum's specimen was taken on strength by the USN as serial 126464 in April 1953 and was stationed at naval air stations at Cherry Point, North Carolina, Jacksonville, Florida, and Norfolk, Virginia, as well as on the carriers USS *Midway*, USS *Forrestal* and USS *Bennington*.

It was obtained by the RCN in August 1957 and, after overhaul and refinishing in RCN colours, was assigned to fighter squadron VF-870, which alternated in location between *Shearwater* and the *Bonaventure*. During its service it was part of the four-plane RCN exhibition team known as the Grey Ghosts.

After its retirement in 1962 the aircraft sat outside at *Shearwater*, stripped of its engines and instruments and other useful equipment. In June 1965 it was donated to the Canadian War Museum and moved to Rockcliffe. No.400 Air Reserve Squadron at Downsview, Ontario, offered to restore it in 1975. The work began in 1976 and engines were found together with much but not all equipment; the restoration was completed in 1984.

McDonnell CF-101B Voodoo

SPECIFICATIONS

Powerplant: Two 16,700 lb (7575 kg) st Pratt & Whitney J57-P-55

Wing span: 38 ft 8 in (11.8 m)

Length: 67 ft 5 in (20.5 m)

Height: 18 ft (5.5 m)

Wing area: 368 sq ft (34.2 m²)

Weight, empty: 28,970 lb (13 141 kg)

Weight, gross: 54,664 lb (24 796 kg)

Speed, max.: 1,134 mph at 35,000 ft (1825 km/h at 10 670 m)

Speed, cruising: 551 mph (887 km/h)

Climb: 49,200 ft/min (15 000 m/min)

Service ceiling: 54,800 ft (16 700 m)

Early in the Korean War the USAF found it needed an escort fighter for its Boeing B-29 bombers and revived the design of the McDonnell XP-88 Voodoo, which had not gone into production. The Voodoo was redesigned with over four times the power of the original and greatly increased range, and was inevitably larger and heavier. The new machine, the F-101A, was ordered directly into production and was first flown on September 29, 1956.

Reconnaissance versions were developed, and work on an all-weather version, the F-101B, began in 1955, with the first flight taking place in March 1957. The type had a long service record in the USAF and established several speed records, including an eastbound coast-to-coast time of 3 hours, 7 minutes, 43 seconds and a westbound time of 3 hours, 36 minutes, 33 seconds.

In Canada the RCAF had been counting on the Avro CF-105 Arrow to replace the all-weather Avro Canada CF-100 in protecting northern areas. The sudden cancellation of the CF-105 contract changed all this, and the F-101B Voodoo was shortly selected to fill the Arrow's intended role.

The first F-101Bs, designated CF-101Bs in Canadian service, arrived in 1961 and the type served well in its all-weather role up to 1984, when the McDonnell CF-18 Hornet entered CF service. Five RCAF/CF squadrons flew Voodoos: No.409, Comox, British Columbia; No.410, Uplands, Ontario; No.414, Namao, Alberta; No.416, Chatham, New Brunswick; and No.425, Bagotville, Quebec. A single specially equipped Voodoo was acquired in 1982 to serve as an electronic counter-measures aircraft. Finished in black, it became known as the "Electronic Voodoo." It seems unfortunate that the opportunity was not taken during the Voodoos' long service here to set some Canadian records that were well within their capability.

The Museum's specimen, 101025, ex-USAF 57-340, was taken on strength in 1971 by the Canadian Forces in exchange for a less well equipped CF-101B, which was returned to the USAF along with 45 others. It served with No.409 Squadron at Comox from March 1972 to June 1980, then with No.416 Squadron at Chatham from December 1980 to July 9, 1984, when it was flown to Uplands. It was then airlifted to Rockcliffe on July 19, less engines, which followed by surface transport.

McDowall Monoplane

Before World War I many young aviation enthusiasts built their own aircraft, sometimes to their own designs and sometimes following the designs of others. Such a young man was Robert McDowall, a civil engineer and land surveyor in Owen Sound, Ontario.

On a visit to England and France in the spring of 1910, McDowall saw heavier-than-air flying machines for the first time and his interest was aroused. He shortly bought a used fan-type Anzani engine in New York and proceeded to design an aircraft to suit it. During the winter of 1911-12, McDowall and a cousin, George A. Ferguson, started construction of the aircraft in a local machine shop. As the work was carried out as a part-time project in the winters only, the machine was not completed until 1915. The aircraft was a shoulder-wing monoplane

SPECIFICATIONS

Powerplant: One 25 hp (fan type)
Anzani
Wing span: 27 ft 6 in (8.4 m)
Length: 21 ft 6 in (6.6 m)
Height: 7 ft 5 in (2.3 m)
Wing area: 139.6 sq ft (13 m²)
No other information is available.

and has often mistakenly been called a Blériot, but in all other respects it differed from the well-known Blériot XI of cross-Channel flight fame.

The machine was taken to a nearby farm and McDowall made several unsuccessful attempts to fly it, although apparently hops of unknown length were made. It was then stored until 1916 or 1917 when two schoolboys attempted to improve the engine's performance and cut away a portion of the trailing edges of the wings with the intention of fitting ailerons. This was never done and the machine was again stored.

About 1920 Edward Pratt of Durham, Ontario, bought the aircraft and used it as an ice scooter on a nearby lake. In 1958 Keith Hopkinson of Goderich, Ontario, bought it from Pratt and it hung from the rafters of his hangar until the spring of 1967, when it was bought from his estate by the National Aviation Museum and brought to Ottawa. It is the oldest surviving Canadian-built aircraft, and although it met with only limited success, it is typical of the efforts of many of the early experimenters.

It was displayed for several years in the Museum of Science and Technology just as it was received, and then stored. Restoration of the aircraft to its original condition as constructed by Robert McDowall was completed in 1987.

Messerschmitt
Me 163B-1a Komet

SPECIFICATIONS

Powerplant: One 1700 kg
              (3,748 lb) st Walter
              HWK 509A-2

Wing span: 9.3 m (30 ft 6 in)

Length: 5.9 m (19 ft 4-1/4 in)

Height (on take-off dolly): 2.8 m
                              (9 ft 2 1/4 in)

Wing area: 18.5 m² (199 sq ft)

Weight, empty: 1905 kg (4,200 lb)

Weight, gross: 4310 kg (9,502 lb)

Speed, max.: 955 km/h (593 mph)

Speed, cruising: 925 km/h (575 mph)
                   approx.

Climb: 4875 m/min (16,000 ft/min)

Service ceiling: 12 190 m (40,000 ft)

The most remarkable aircraft in concept and performance to operate in World War II was the Me 163 rocket fighter, the only rocket fighter ever to enter service. The design originated with the work carried out on tailless aircraft by the respected scientist Dr. Alexander Lippisch, which began in 1921 with glider designs.

The Me 163 started at the German Research Institute for Sailplanes as the DFS 194, intended as a piston-engined test vehicle for a future rocket-powered fighter. In 1939 Dr. Lippisch and the project were transferred to the Messerschmitt Co., and the DFS 194, now rocket-powered, first flew in August 1940. Work then started on the Me 163, with gliding trials on the prototype starting in the spring of 1941 and powered flights in August. The operational version, the Me 163B, although very similar to the original version, was a completely new design. The first trial of a pre-production Me 163B took place in August 1943.

The Me 163 was designed to have performance far exceeding that of contemporary conventional aircraft, although it was of short range and duration. The aircraft was armed with two 30mm cannons. The aircraft was unusual in that it took off from a dolly and landed on a skid on the underside of the fuselage. Its in-flight performance and handling were good, but it was difficult on take-off and its fuel, *T-stoff* and *C-stoff*, had to be handled very carefully and could be dangerous. It glided to a landing after its fuel for 7.5 minutes of rocket power was exhausted, and so, regardless of weather or other conditions, each landing had to be made perfectly as there was no possibility of going around again.

The first Me 163 operational units were formed in the spring of 1944 and the first contacts with Allied aircraft were made in May, but it was not until June that Me 163s were assembled in numbers. In combat the type was not as effective as might be imagined. At first it suffered from the ground controllers not knowing the best way to handle the fast Me 163, but, more importantly, the fast closing speed of the aircraft on its targets combined with the slow rate of fire of its cannons meant that the pilot had to be an exceptionally good marksman or very lucky to score any hits. Only 339 Me 163s were delivered.

The Museum's two specimens were brought to Canada at the end of the war and one, serial 191916, was stored at Calgary and brought to Rockcliffe on November 15, 1964. It was refinished in the markings of 1 JG/400 of the Luftwaffe by the Canadian War Museum and put on display there. The other was rescued in late 1957 by the Canadian War Museum after being displayed outside at RCAF Station St. Jean, Quebec. It had deteriorated and was restored during 1974-78 by the National Aviation Museum in its original factory paint scheme. It was on loan to the USAF Museum at Wright-Patterson Air Force Base from 1978 to 1985.

Nieuport 12

SPECIFICATIONS

Powerplant: One 110 hp Clerget 9Z
Wing span, upper: 9 m (29 ft 6-3/8 in)
Wing span, lower: 7.4 m (24 ft 3-3/8 in)
Length: 7 m (23 ft)
Height: 2.7 m (8 ft 10 1/4 in)
Wing area: 22 m² (237 sq ft)
Weight, empty: 550 kg (1,213 lb)
Weight, gross: 850 kg (1,874 lb)
Speed, max.: 146 km/h at 2000 m
(91 mph at 6,560 ft)
Climb: 1000 m (3,280 ft)
in 5 min 40 sec
Ceiling: 4000 m (13,120 ft)

Édouard de Niéport flew his first aircraft in 1909 and established a factory, Société anonyme des Établissements Nieuport. His various monoplane types were noted for their speed and set many records before World War I. They were adopted by the air services of a number of countries and used by civil pilots.

In 1914 the Nieuport company introduced a sesquiplane, their Model 10, powered with an 80 hp Gnome engine. This was a two-seater and established the configuration that was characteristic of almost all the famous Nieuport types of World War I.

In 1915 the Nieuport company introduced a slightly larger two-seater, the Nieuport 12, powered by the more powerful 110 hp Clerget. This type saw service with several of the Allies, including France, Italy, Russia, and both the Royal Flying Corps and Royal Naval Air Service of Great Britain. They were usually armed with one Lewis gun for the observer, but a few British machines were converted to single-seaters with a machine gun over the upper wing for the pilot, and at least one British machine had guns fitted for both pilot and observer.

In service the Nieuport 12 did not distinguish itself and was unpopular with its crews, and at least one report severely criticized the type. It went into service with the RFC and RNAS in 1916, but all were removed from front line service by the spring of 1917 to soldier on for a few months more on training duties.

The Museum's specimen is a French-built example (some were built in England by William Beardmore and Co. Ltd.). It has a fin and rudder unlike those on any other known Nieuport 12; how or why this came about is not known. Beyond this, nothing definite is known of its early history and of its coming to Canada. The earliest known reference to it occurs in a November 1936 file of the Aeronautical Museum, which states that the War Trophies Board was looking for storage for "a Nieuport reported to have been donated by the French Government." In 1956 it was recorded that most of the fabric was missing and there were no markings on the machine.

The aircraft was restored in 1958 by 6 Repair Depot, RCAF, and finished in spurious colours, markings and serial number, and the lower wing fittings are such that it cannot be correctly rigged. It has not been on Museum display and will be restored in due course, probably in RNAS markings as many Canadians were associated with the type in RNAS service.

Nieuport 17

SPECIFICATIONS

Powerplant: One 110 hp
Le Rhône JB

Wing span, upper: 8.16 m
(26 ft 9-1/4 in)

Wing span, lower: 7.8 m (25 ft 6 in)

Length: 5.8 m (19 ft)

Height: 2.4 m (7 ft 10-1/2 in)

Wing area: 14.75 m² (159 sq ft)

Weight, empty: 375 kg (827 lb)

Weight, gross: 560 kg (1,235 lb)

Speed, max.: 165 km/h at 2000 m
(102 mph at 6,560 ft)

Climb: 2000 m (6,560 ft)
in 6 min 50 sec

Service ceiling: 5300 m (17,390 ft)

As related under the Nieuport 12, the Nieuport family of sesquiplanes began in 1914 with the Model 10 two-seater. In 1915 the diminutive Nieuport 11 was designed, a single-seat version refined and developed from the Model 10, powered by an 80 hp Le Rhône engine and armed with a single Lewis gun over the upper wing. Its size quickly earned it the sobriquet "le bébé Nieuport," a term which was confusingly applied to later Nieuports. It quickly established a name for itself, was adopted by several of the Allies and was soon followed by the almost identical but more powerful 110 hp Nieuport 16.

The Nieuport 16 was quickly followed by the somewhat larger Nieuport 17, which was adopted by all the Allies and copied by the Germans and became one of the classic fighters of World War I. The standard version had a single Vickers gun firing through the propeller, but the British machines and some others had a single Lewis gun over the upper wing. Eight Le Prieur rockets could be carried on the wing struts. While still other Nieuport models of somewhat similar appearance followed, it was the Model 17 that firmly engraved the Nieuport name in the annals of war in the air.

The French pilots, as might be expected, were the virtuosos of the Nieuport, and all the great French aces of the period, like Nungesser, Guynemer, Dorme and others, made their name with the type. Six RFC squadrons and eight RNAS squadrons used the Nieuport 17, but possibly No.60 Squadron RFC was its leading British exponent. Canada's first aerial VC winner, Lt. W.A. Bishop of No.60 Squadron, flew a Nieuport 17 on his single-handed raid on a German aerodrome in which he destroyed three enemy aircraft. Of course other Canadians, like Lts. A.D. Bell-Irving and W.W. Rogers, flew the Nieuport 17, but Billy Bishop's name will always be associated with the type in British service.

The Museum's specimen is a reproduction completed in 1962 by Carl R. Swanson at Sycamore, Illinois, and was the first of a number of fine World War I aircraft he has made. It was finished originally to represent a French machine of Escadrille N.124 but was changed to represent Billy Bishop's machine of No.60 Squadron RFC. It was first flown by W/C P.A. Hartman at Rockcliffe on May 4, 1967. It has subsequently been used by the Museum for flying displays at several Canadian centres.

✳

Noorduyn UC-64A Norseman VI

SPECIFICATIONS

Powerplant: One 600 hp Pratt & Whitney R1340-AN-1
Wing span: 51 ft 6 in (15.7 m)
Length: 32 ft (9.8 m)
Height: 10 ft 3 in (3.1 m)
Wing area: 325 sq ft (30.2 m²)
Weight, empty: 4,680 lb (2123 kg)
Weight, gross: 7,400 lb (3357 kg)
Speed, max.: 162 mph (261 km/h)
Speed, cruising: 148 mph (238 km/h)
Climb: 5,000 ft (1520 m) in 6.5 min
Service ceiling: 17,000 ft (5180 m)

R.B.C. Noorduyn was born of British-Dutch parents and commenced his long and notable career in aviation with the Sopwith Aviation Co. in England in 1913. In the 1920s he came to the United States, where he worked for Fokker and later Bellanca, and it was with these two companies that he first came into contact with Canadian aviation and particularly bush flying.

Late in 1934 Noorduyn came to Montreal and with Canadian financial backing began the design of a new bush aircraft. This was completed during the winter of 1934-35 and construction began at Cartierville in the spring. Long hours of work by Noorduyn's personnel resulted in the 420 hp Wright Whirlwind-powered prototype Norseman flying on November 14. The first production version, the Norseman II, also Whirlwind-powered, followed in 1936. However, the type was underpowered and it was only with the installation of the 550 hp Pratt & Whitney Wasp S3H1 in the Norseman IV that the Norseman really came into its own. The Norseman IV prototype was first flown on November 5, 1936. This version became popular with Canadian bush operators and was also ordered by the RCAF.

After the United States entered World War II, the Norseman was adopted by the USAAF and ordered in quantity as a utility aircraft. The USAAF version was modified slightly to suit their military requirements and designated by them as the UC-64A and by Noorduyn as the Norseman VI. The final production Norseman development was a postwar civil version, the Norseman V, which may be considered as a civilianized Norseman VI. The last Norseman was completed in December 1959, just over 24 years after the prototype was first flown. The Norseman was the most numerous of all Canadian-designed aircraft up to that time, with 904 built.

Some Norsemans still remain in service, and throughout the type's career it has enjoyed an excellent reputation as a workhorse in civil and military use. Following the war, surplus military Norsemans entered civil use in many countries and with the air services of Australia, the Netherlands East Indies, Indonesia, Norway and Sweden.

The Museum's specimen is a UC-64A Norseman VI, started as USAAF 43-5145 but completed as RCAF 787, the RCAF's first Mk.VI, and accepted on June 10, 1943. It served during the war with No.1 Wireless School at Montreal. In the postwar period it served with Nos. 412 and 408 Squadrons at Rockcliffe and spent a final year with No.121 Communications and Rescue Flight before being stored in 1954. It was transferred to the RCAF's historic aircraft collection at Rockcliffe in 1964.

North American AT-6 Harvard II and T-6J Harvard 4

SPECIFICATIONS

Powerplant: One 600 hp Pratt &
    Whitney Wasp
    R1340 AN-1

Wing span: 42 ft-1/2 in (12.8 m)

Length: 28 ft 11 in (8.8 m)

Height: 11 ft 8-1/2 in (3.6 m)

Wing area: 253.7 sq ft (23.6 m²)

Weight, empty: 3,995 lb (1812 kg)

Weight, gross: 5,235 lb (2375 kg)

Speed, max.: 180 mph at 5,000 ft
    (290 km/h at 1524 m)

Speed, cruising: 140 mph at 12,800 ft
    (225 km/h at 3900 m)

Initial climb: 1,300 ft/min
    (396 m/min)

Service ceiling: 22,000 ft (6710 m)

The Harvard must be considered one of the outstanding aircraft of World War II, and there were very few American or British Commonwealth single-engine pilots that did not receive their advanced training in it. Following the war, it was adopted by many other countries.

The prototype of the Harvard was the NA-16 developed by North American Aviation Inc. of Inglewood, California, to meet the 1937 requirements of the USAAC for a single-engine service trainer to suit the new combat aircraft then coming into use. It won the competition and was placed in production. By 1939 it had been developed into the AT-6 in American designation and the Harvard II in the British Commonwealth.

The RCAF had just selected the Harvard as its first service trainer when the war broke out, and the aircraft was adopted for use in the British Commonwealth Air Training Plan. The first RCAF Harvards were supplied by North American, but in 1940 the RCAF placed an order with Noorduyn Aviation Ltd. of Montreal. The company went on to produce 2,800 Harvards for the RAF and RCAF during the war. In the early 1950s Canadian Car & Foundry at Fort William, Ontario, produced 550 Harvard 4s, US designation T-6J, a radio trainer version. This made the Harvard the most numerous of all Canadian-made aircraft.

In service the Harvard was universally liked and respected and proved to be just the right machine for making the transition from elementary to service types of aircraft. In the postwar years many Harvards appeared on the civil aircraft register, mostly flown for pleasure by their owners and some of them marked in their wartime finish.

The Museum holds three specimens, two Harvard IIs and one

Harvard 4. Both Mk.IIs were built by North American. Serial 2532 was accepted on August 17, 1940, and served in several training schools until it was stored in 1962 and then transferred to the RCAF historic aircraft collection at Rockcliffe in 1964. Serial 3840 was accepted on May 17, 1941, and also served at several training schools until it became a ground instructional airframe at Camp Borden and then was given to the National Aviation Museum in November 1962. Serial 3840 was displayed on a pedestal outside the Canadian War Museum from 1964 to 1983 and is now weathered. The Mk.4 was accepted on September 5, 1952, and after instructional use was transferred to the RCAF historical aircraft collection in 1964.

North American P-51D Mustang IV

SPECIFICATIONS

Powerplant: One 1,680 hp
Rolls-Royce
(Packard) Merlin 69
Wing span: 37 ft-1/4 in (11.3 m)
Length: 32 ft 3 in (9.8 m)
Height: 8 ft 8 in (2.7 m)
Wing area: 235 sq ft (21.8 m²)
Weight, empty: 7,000 lb (3175 kg)
Weight, gross: 9,200 lb (4173 kg)
Speed, max.: 442 mph at 24,500 ft
(711 km/h at 7468 m)
Climb: 20,000 ft (6100 m) in 10 min
Service ceiling: 42,500 ft (12 950 m)

The prototype of the P-51 was designed and built in 117 days in 1940 to fill a British requirement for a low-altitude fighter powered with the Allison V-1710-39 engine. The P-51 entered production for Britain and shortly was adopted by the USAAF, and it was successful in its low-altitude role. Adapting it to take the Packard-built Rolls-Royce Merlin engine resulted in a fine high-altitude fighter, probably the best of the war, as it had not only excellent speed and climb but, with drop tanks, a radius of action of 850 miles (1370 km), far exceeding that of its contemporaries. The first Merlin-powered P-51 was the P-51B/51C

Mustang III, which was superseded by the P-51D Mustang IV with its bubble canopy for improved pilot vision.

The first Mustang Is went into action with the RAF in the summer of 1942 and operated extensively and successfully in Tunisia and the Mediterranean area. The first Mustang IIIs entered service use in February 1944 and escorted the USAAF daylight bombing raids into Germany. This was the first time fighter escorts were available, and the previously heavy bomber losses were sharply reduced.

Three RCAF squadrons, Nos. 400, 414 and 430, flew the low-altitude Mustang Is from bases in Britain from June 1942 to December 1944. Flying the high-altitude Mustang III from March to August 1945, No.442 Squadron carried out escort duties over Europe. No.441 Squadron also briefly flew the Mustang III in 1945 but not operationally.

The RCAF adopted the P-51D Mustang IV in 1947 and took delivery of 130. Two regular RCAF squadrons, Nos. 416 and 417, were equipped with them, as were six auxiliary squadrons, Nos. 402, 403, 420, 424, 442 and 443. The Mustang IVs were retired from squadron use in 1956 but a few soldiered on in training units until 1960.

The Museum's specimen was built in 1944 at North American's Inglewood plant as USAAF 44-73347. Its American history is unknown but it was taken on strength by the RCAF in March 1951 and it served with No. 442 City of Vancouver (Aux.) Squadron but is now finished as a Mustang III of No.442 Squadron in 1944.

North American TB-25L Mitchell 3PT

The B-25 Mitchell medium bomber proved a most effective weapon in the hands of the Allies. It served on all fronts and was used for a variety of purposes. A torpedo was sometimes carried externally for attacks against Japanese shipping, while a 75mm field gun was mounted in the B-25G and B-25H models also for attacks against shipping.

In spite of its widespread and effective deployment on all fronts, what the B-25 is undoubtedly best remembered for is the courageous Tokyo, Kobe, Yokohama and Nagoya bombing raid in April 1942. The attack was led by Lt.-Col. J.H. Doolittle and was carried out by 16 B-25Bs which took off from the USS *Hornet* some 800 miles (730 km) from Japan. After bombing their targets, the aircraft flew on to China, where they crash landed. The raid demonstrated to Japan that it was not immune to air attack about four months after the attack on Pearl Harbor.

The B-25 stemmed from a private-venture design by North American which first flew in June 1939. Based on its performance, the USAAC requested that development be continued with some changes.

SPECIFICATIONS

Powerplant: Two 1,700 hp
            Wright R-2600-13
Wing span: 67 ft 7 in (20.6 m)
Length: 52 ft 11 in (16.1 m)
Height: 16 ft 4 in (5 m)
Wing area: 610 sq ft (56.7 m²)
Weight, empty: 19,418 lb (8808 kg)
Weight, gross: 35,000 lb (15 876 kg)
Speed, max.: 272 mph at 13,000 ft
            (438 km/h at 3960 m)
Speed, cruising: 230 mph (370 km/h)
Climb: 15,000 ft (4570 m) in 19 min
Service ceiling: 24,200 ft (7380 m)

The resulting machine was the B-25, the prototype of which first flew on August 19, 1940. It was named after Col. William Mitchell, the well known proponent of air power.

The RCAF did not use the B-25 on operations against the enemy during the war. However, No.13 Squadron (later 413) was reformed at Rockcliffe in 1943 to do photographic research and used B-25D Mitchell IIs from 1944 to 1947. No.418 (Auxiliary) Squadron at Edmonton used the Mitchell II and 3 from January 1947 to March 1959 in its light bomber role. No.406 (Auxiliary) Squadron used Mitchell 3s at Saskatoon from June 1948 to June 1958, also in a light bomber role. No.412 Squadron at Rockcliffe used the Mitchell 3 from September 1956 to November 1960 in their transport role, the aircraft having been converted for this purpose.

The Museum's specimen was built at Kansas City as a B-25J 44-86699 and converted to a pilot trainer, TB-25L, in the postwar period. It was acquired by the RCAF in January 1952 and served at 1 Flying Instructors School, Trenton, until 1959. It was then briefly at 5 Air Observer School Winnipeg before going to the Central Experimental and Proving Establishment at Uplands. It was stored in 1962 and transferred to the RCAF historical aircraft collection in 1964 in the markings of an aircraft of No.98 Squadron (RAF).

North American *F-86E Sabre 6*

SPECIFICATIONS

Powerplant: One 7,275 lb
      (3299 kg) st Avro
      Canada Orenda 14

Wing span: 37 ft 11-1/2 in (11.6 m)

Length: 37 ft 6 in (11.4 m)

Height: 14 ft 9 in (4.5 m)

Wing area: 302.3 sq ft (28.1 m²)

Weight, empty: 10,618 lb (4816 kg)

Weight, gross: 14,613 lb (6628 kg)

Speed, max.: 606 mph (975 km/h)
      at sea level

Speed, cruising: 489 mph at 45,000 ft
      (787 km/h at 13 720 m)

Climb: 35,000 ft
      (10 670 m) in 4 min 42 sec

Service ceiling: 54,000 ft (16 460 m)

The F-86 Sabre was the leading NATO fighter of its time and was used initially by the American, British, Canadian, and West German air forces and later by others as well. It was a logical day fighter choice for the RCAF, which gave Canadair Ltd. of Montreal a contract to build the type. The first Canadair Sabre was flown in August 1950.

The Sabre was developed from a straight-wing design; the first prototype to fly in swept-wing form was the XP-86 on October 1, 1947. Over the production life of the Sabre, a number of variants were built with modifications to equipment and airframe. All were similar in appearance except for the F-86D and F-86L, which had all-weather radar in the nose and a chin air intake and were increased in length.

The early Canadair-built Sabres, the Mks.1, 2 and 4, were very similar to their American counterparts but the two later Canadair variants, the Sabre 5 and 6, had Avro Canada Orenda 10 and 14 engines respectively, and with the extra power they were the best of their excellent family. The first Orenda installation was done by North American at Canada's request. The sole Canadair-built Mk.3, first flown on June 4, 1952, was used in 1953 by Jacqueline Cochran to set three international records at Edwards Air Force Base, California. The two Orenda-powered production versions, the Mks.5 and 6, followed shortly.

Canadair-built Sabres were supplied originally to the air forces of Britain, Canada, Columbia, West Germany and South Africa, and, later, RCAF machines were passed on to still others under the Mutual Aid Program. The RCAF Sabre 2s were the first swept-wing aircraft with the NATO forces in Europe and the Sabre 4s were the first swept-wing aircraft in the RAF.

In the RCAF, Sabres served in Canada and with NATO forces in Europe. The Sabre 2 began its RCAF service in 1951 and was phased out in 1954 and 1955. Most Sabre 4s went directly to the RAF, with which the majority served in West Germany, but a few served briefly with the RCAF before going on to the RAF. The Sabre 5 started its RCAF career in 1955 and was followed by the more powerful Mk.6 in 1965. Sabres were greatly liked by their pilots, who regarded it as "a real pilot's aircraft." The Canadian public best remembers the Sabre as the mount of the RCAF Golden Hawks aerobatic team, who gave excellent exhibitions from coast to coast.

The Museum holds two specimens, both Sabre 6s. One, RCAF 23455, is in the markings of No.444 (Cobra) Squadron, 1 Air Division, and the other, RCAF 23651, is a Golden Hawk machine. The skinning on one side of the latter has been removed to show the interior.

Piasecki (PV-18) HUP-3

SPECIFICATIONS

Powerplant: One 550 hp
        Continental R-975-46
Length (fuselage only): 32 ft (9.8 m)
Height: 12 ft 6 in (3.8 m)
Rotor diameter: 35 ft (10.7 m)
Weight, empty: 4,132 lb (1874 kg)
Weight, gross: 5,750 lb (2608 kg)
Speed, max.: 100 mph (161 km/h)
Speed, cruising: 80 mph (129 km/h)
Climb, inclined: 1,000 ft/min
        (305 m/min)
Absolute ceiling: 10,000 ft (3050 m)

Frank N. Piasecki formed the PV Engineering Forum in 1943, which designed and built the successful PV-2 single-rotor helicopter that year. On this type Piasecki became the first American to obtain a helicopter pilot's licence.

Their next helicopter was the PV-3, which featured tandem rotors, the first production machine of this type. It went into service with the US Navy as the HRP-1 Rescuer but was widely known as the "Flying Banana" owing to its appearance. The tandem rotor offered a wider choice of centre of gravity location with a small frontal area and was an appreciable advance over contemporary helicopter designs.

The PV-18 model was developed to suit a USN requirement for a carrier-based utility helicopter for search and rescue operations and general transportation use. This six-place machine was accepted by the USN under their HUP-3 designation and the US Army under their H-25 designation. In the navy the type was named the Retriever and in the army it was called the Army Mule. Like its predecessor, the PV-18 had tandem rotors which could be folded for shipboard storage and which were interconnected by shafts to an engine mounted in the rear of the fuselage. A large floor hatch was provided through which loads of up to 400 lb (180 kg) could be hoisted by a winch. The HUP-3 and H-25A had strengthened floors for heavy loads, stretcher facilities and hydraulic boost on the controls, all of which made it an appreciable advance over earlier types. Production ended in 1954 after 339 were built, with the majority going to the USA and lesser numbers to the US Army, France's Aéronavale and the Royal Canadian Navy.

As the Piasecki Helicopter Corp., which had evolved from the PV Engineering Forum in 1946, was reorganized as the Vertol (for vertical take off and landing) Corp., the type is sometimes referred to as the Vertol HUP-3. This is incorrect as all machines of the type had been completed before the name change.

The RCN obtained three HUP-3s in 1954. All RCN HUP-3s served with helicopter squadron HU-21. In Canada the name Retriever was not used for the type, which flew from the land base HMCS *Shearwater*, RCN carriers and occasionally survey ships. It was used for search and rescue work along with some survey and general transportation assignments.

The Museum's specimen served with helicopter squadron HU-21 until it was retired ten years later. It was bought by the Canadian War Museum on February 24, 1965.

Pitcairn-Cierva
PCA-2/PA-21 *Autogiro*

SPECIFICATIONS

Powerplant: One 420 hp Wright
R-975-E2 Whirlwind
Wing span: 30 ft (9.1 m)
Rotor diameter: 45 ft (13.7 m)
Length: 23 ft 1 in (7 m)
Height: 13 ft (4 m)
Wing area: 88 sq ft (8.2 sq m)
Rotor blade area: 159.9 sq ft
(14.9 m²)
Weight, empty: 2,029 lb (920 kg)
Weight, gross: 3,000 lb (1360 kg)
Speed, max.*: 118 mph (190 km/h)
Speed, cruising*: 98 mph (158 km/h)
Climb*: 800 ft/min (244 m/min)
Service ceiling*: 15,000 ft (4570 m)

The Autogiro was developed by Juan de la Cierva in Spain. His work started in 1920 but it was not until January 17, 1923, that the first successful Autogiro, the fifth prototype, was flown. The Autogiro's rotor is not engine-driven like that of a helicopter but rotates by the action of the air stream. Consequently the Autogiro cannot hover, but it can descend almost vertically and land with about a 15-foot run, and it can take off with a short run and climb steeply but not vertically. The development of the Autogiro inevitably lead to the successful helicopter in 1936.

Autogiro development was transferred to England in 1926 and continued by the Cierva Autogiro Co. In 1929 Harold F. Pitcairn obtained the North American rights to the Autogiro and formed the Pitcairn-Cierva Autogiro Co. at Willow Grove, Pennsylvania. The company designed and built 15 types of Autogiros. After completing the experimental PCA-1 model in 1929, Pitcairn flew the prototype PCA-2 in the spring of 1930. The three-place PCA-2 became, in April 1931, the first rotary-wing aircraft certified for commercial use in the United States. 24 PCA-2s were built, including three for the US Navy designated XOPs and used for experimental observation work. Autogiros were used for the many purposes in which their almost-vertical take-off and landing capability could be put to use, including crop dusting, news gathering, advertising and promotional work, and executive transport in restricted areas.

In Canada the Fairchild Aircraft Co. of Longueuil, Quebec, had the sales agency and imported a PCA-2, registered CF-ARO, the first rotary-winged aircraft in Canada. It was also the first Autogiro to

* Performance figures are given with the 300 hp Wright R-975 engine. Improved figures with the 420 hp Wright R-975-E2 engine are not available.

be looped – by Fairchild's pilot Godfrey Dean at Willow Grove in September 1931. Its unusual flight characteristics made it a star performer at the many stops of the Trans Canada Air Pageant during the summer of 1931.

The Museum's specimen was delivered to the Standard Oil Co. of New York (Socony) in March 1931, registered NR-26. It was used for test purposes before being sold the following November to the Sealed Products Corp., makers of piston rings, who used it for test and advertising purposes. In 1932 it made a 60,000-mile flight to visit 225 cities in 22 states and Mexico, promoting the company's products. During this flight pilot Don Walker made 2,100 landings and carried 3,066 passengers. Its subsequent career is unknown at present but sometime during the 1930s its original 300 hp Wright R-975 engine was replaced by a 420 hp R-975-E2, changing its model designation from a PCA-2 to a PA-21.

Royal Aircraft Factory B.E.2c

SPECIFICATIONS

Powerplant: One 90 hp R.A.F. 1a

Wing span, upper and lower: 36 ft 10 in (11.2 m)

Length: 27 ft 3 in (8.3 m)

Height: 11 ft 4 in (3.5 m)

Wing area: 396 sq ft (36.8 m²)

Weight, empty: 1,370 lb (621 kg)

Weight, gross: 2,142 lb (972 kg)

Speed, max.: 72 mph at 10,000 ft (116 km/h at 3050 m)

Climb: 6,500 ft (1980 m) in 20 min

Service ceiling: 10,000 ft (3050 m)

The prototype of the B.E. series, the B.E.1, was first flown in December 1911. The type then went into production as the B.E.2, and this version was subsequently modified and went into use in various versions designated B.E.2a through to B.E.2e and sub-types B.E.2f and 2g. The prototype B.E.2c was first flown on May 30, 1914.

The B.E.2c was the most numerous and widely used British aircraft of the early World War I period. Its main duties were as a two-seat general reconnaissance and artillery observation machine, and secondary uses were as a single-seat bomber and a Home Defence fighter. The type was much and unjustly maligned by opponents of the Royal Aircraft Factory and its products, and unfortunately their accusations have been perpetuated. In fact the B.E.2c was a very good machine for its time, but late in its career it had to soldier on, like some other types, longer than desirable due to delays in introducing its successors.

In its Home Defence role it became most successful. Flown as a single-seater to improve its high-altitude performance and armed with some or all of the following, Lewis gun, Le Prieur rockets and Ranken darts, it proved the most effective aircraft against the German dirigibles. The stability of the B.E.2c greatly aided the night-flying pilots and six dirigibles were victims of the B.E.2c. It has been said that its success in this role alone justified its existence.

The Museum's specimen was built by the British & Colonial Aeroplane Co. at Bristol and issued to No.39 Squadron on Home Defence duties. The aircraft, serial 4112, was a single-seater armed with a Lewis gun. On the night of September 23/24 1916, Lt. F. Sowery intercepted Zeppelin L.32 in 4112 and shot it down. Subsequently 4112 was converted to the usual two-seat standard, the armament was removed and an enlarged fin was fitted.

The aircraft was brought to Canada less engine and propeller in 1919 and exhibited at Toronto. It was then stored until the late 1930s, when it was exhibited at the Aeronautical Museum at the National Research Council until 1940. It was stored until 1957, when it was restored by the RCAF. An engine was donated by the South Kensington Science Museum in 1958. Its original propeller was donated in 1962 at the suggestion of W/C F.B. Sowery, son of Lt. F. Sowery.

The Sikorsky Aircraft Division of United Aircraft Corp. developed the first production helicopter, the R-4, the prototype of which first flew in January 1942. A more powerful observation helicopter, the R-5, followed. The S-51 was an early postwar development of the R-5 originally intended as a four-place machine for the commercial market. It first flew in February 1946 and soon became the first Sikorsky helicopter to receive approval for civil use. As it happened, most S-51s went into military service.

Sikorsky made 214 S-51s, and a further 165 were made by Westland Aircraft Ltd. in England in two versions.

In service use, the S-51, under various military designations,

Sikorsky (S-51) H-5 Dragonfly

SPECIFICATIONS

Powerplant: One 450 hp Pratt &
        Whitney R-985-AN-5
        Wasp Jr.

Rotor diameter: 49 ft (14.9 m)

Length: 57 ft 8 in (17.6 m)

Height: 12 ft 11 in (3.9 m)

Weight, empty: 3,810 lb (1728 kg)

Weight, gross: 5,500 lb (2495 kg)

Speed, max.: 103 mph (166 km/h)

Speed, cruising: 85 mph (137 km/h)

Vertical climb: 200 ft/min (61 m/min)

Inclined climb: 1,000 ft/min
        (305 m/min)

*Hovering ceiling without
ground effect:* 3,100 ft (945 m)

Service ceiling: 13,500 ft (4115 m)

appeared with the following air services: RAAF, RAF, RCAF, US Army, USAF, USMC and USN. Most services used the type on search and rescue duties; the RAF used it for casualty evacuation in Malaya in the 1950s; the USMC and USN used it for a variety of purposes in the Korean War; and two S-51s accompanied Admiral R.E. Byrd on his 1946-47 Antarctic expedition.

The RCAF's seven S-51s, or H-5s as they were designated by the RCAF and USAF, were taken on strength in April 1947 and were the first RCAF helicopters. As would be expected, they served to indoctrinate RCAF personnel in the flying and maintenance of this new type of flying machine, giving first-hand experience in the capability and restrictions of the helicopter and, most important, their performance under Canadian winter conditions. Consequently the H-5s spent most of their lives with training or experimental units, as well as doing some search and rescue work.

The Museum's specimen, RCAF 9601, was taken on strength on April 5, 1947, the first of the seven in the RCAF. It was assigned to Trenton on training duties, then to Edmonton, then back to Trenton in 1950 for repair and overhaul, then to Rivers, Manitoba, for service with the Canadian Joint Air Training School for three years, then on to the experimental unit at Cold Lake, Alberta, for five years, followed by almost five years at Chatham, New Brunswick, before it was handed over to the Museum on January 21, 1965. Two other RCAF H-5s are held by museums, 9602 by the New England Air Museum and 9607 by the Calgary Museum.

Sikorsky (S-55)HO4S-3

SPECIFICATIONS

Powerplant: One 600 hp
        Wright R-1300-3D
Rotor diameter: 53 ft (16.2 m)
Length: 42 ft 2 in (12.9 m)
Height: 14 ft 8 in (4.5 m)
Weight, empty: 5,250 lb (2380 kg)
Weight, gross: 7,540 lb (3420 kg)
Speed, max.: 112 mph (180 km/h)
Speed, cruising: 91 mph (148 km/h)
Climb: 1,020 ft/min (311 m/min)
Vertical climb at sea level: 100 ft/min
                    (30 m/min)

*Hovering ceiling without
ground effect:* 2,300 ft (700 m)

The S-55 prototype first flew on November 10, 1949, and a total of 1,812 were built by Sikorsky and its licensees. The S-55 carried a pilot, co-pilot and ten military passengers or seven commercial passengers. The military and naval forces of 36 countries were the prime users of the type, which first went into service in 1950, but 179 were sold to commercial users including New York Airways, Los Angeles Airways and Okanagan Air Services of Vancouver.

The type was made in several variants with distinguishing model designations but only three were given names: the US Army's H-19 Chickasaw, the Westland-built S-55 Whirlwind, and the French S-55 *Joyeux Éléphant* or "Joyful Elephant." The RCAF version was the H-9 and the RCN the HO4S-3, and both were un-named in their service. The S-55 was the first helicopter to fly the Atlantic, two USAF H-19As flying from Westover Air Force Base, Maine, to Prestwick, Scotland, via Goose Bay, Greenland and Iceland and taking from July 15 to 31, 1952, for the trip because of bad weather. The first helicopter mail service was commenced by two S-55s of Los Angeles Airways in June 1952.

The type was introduced into Canada in December 1952 by the RCN, which bought 13, and the RCAF followed in 1954 with 15. The

RCN used the HO4S-3 for a variety of duties. Fitted with sonar, the type was used for submarine detection. Armed with a homing torpedo, it could also attack naval targets. The HO4S-3 was widely used for rescue work and one hovered continually on guard against a ditching during flying exercises from carriers at sea.

During much of its life the Museum's specimen was stationed at HMCS *Shearwater*, Nova Scotia, but it also served on the carriers *Magnificent* and *Bonaventure*. During this time it carried out eight rescue missions and earned the name "Shearwater Angel." The most famous of these missions occurred on November 17, 1955, when it rescued 21 crew members of the Liberian freighter *Kismet II*, along with the ship's dog and cat. The ship had grounded at the base of a cliff on Cape Breton Island and four rescue flights were made in high winds to remove the crew. The two pilots received the George Medal and the two crew members the Queen's Commendation for the difficult rescue. Other rescues included a man overboard from the *Magnificent*, the same pilot rescued twice in three days following ditching from the *Magnificent*, and two men trapped below a dam in danger of collapsing. The Shearwater Angel was retired to the Museum on May 21, 1970.

Sopwith Pup

The Sopwith Pup was the first of the famous Sopwith single-seat fighters that contributed so much to the Royal Flying Corps' success in the first air war. The prototype first flew in February 1916, and someone who saw it with the two-seater Sopwith 1 1/2 Strutter remarked that it appeared the larger machine had had a pup, so Pup it became. The type was ordered into production for both the RFC and the RNAS.

In service the type became popular because of its excellent flying characteristics, and long after it was superseded on operations it was deemed the most delightful of flying machines. The Pup first appeared on operations in the fall of 1916. Pups were used first in France by the RNAS, with Nos. 3,4,8,9 and 12 (Naval) Squadrons together with No.1 (Naval) Wing all operating Pups there. The RNAS also used the type on shipboard operations. RFC squadrons Nos. 46, 54 and 66 used the type in France, and several RFC Home Defence squadrons used the Pup.

Many Canadians flew Pups on operations and later many more flew it in training. Among noteworthy Canadian Pup pilots was Torontonian F/S/L E.R. Grange of No.8 (Naval) Squadron, who became an ace when he was credited with his fifth enemy aircraft on January 4, 1917; all his kills were made while flying Pups. Other noteworthy Canadian Pup

SPECIFICATIONS

Powerplant: One 80 hp
         Le Rhône 9C
Wing span,
upper and lower: 26 ft 6 in (8.1 m)
Length: 19 ft 3-3/4 in (5.9 m)
Height: 9 ft 5 in (2.9 m)
Wing area: 254 sq ft (23.6 m²)
Weight, empty: 787 lb (357 kg)
Weight, gross: 1,225 lb (556 kg)
Speed, max.: 111 mph (179 km/h)
Climb: 5,000 ft (1525 m)
         in 6 min 25 sec
Service ceiling: 17,500 ft (5335 m)

pilots were Flt. Lieut. L.S. Breadner of Ottawa of No.3 (Naval) Squadron, F/S/L J.J. Malone of Regina also with No.3 (Naval), and F/S/L A.M. Shook of Tioga, Ontario, of No.4 (Naval) Squadron.

The Museum's specimen is a reproduction made with meticulous care by George Neal of Toronto. All work was done by Neal personally except for assistance on the fabric work by former de Havilland Canada fabric workers. The manufacture of the Pup stretched out over seven years before it was first flown at Maple, Ontario, by Neal on September 2, 1967. The aircraft appeared in numerous air shows and fly-ins in the Toronto area, always flown by its maker, until sold to the Museum in October 1973. It was then recovered and finished as Pup B2167, aircraft "L" of 66 Squadron, RFC. It is still flown by Neal for the Museum and has appeared at air shows at several Canadian locations. For civil identification it bears the appropriate registration of CF-RFC.

Sopwith Triplane

SPECIFICATIONS

Powerplant: One 130 hp Clerget 9B
Wing span: 26 ft 6 in (8.1 m)
Length: 18 ft 10 in (5.7 m)
Height: 10 ft 6 in (3.2 m)
Wing area: 231 sq ft (21.5 m²)
Weight, empty: 1,101 lb (499 kg)
Weight, gross: 1,541 lb (699 kg)
Speed, max.: 117 mph at 5,000 ft
         (188 km/h at 1520 m)
Climb: 1,000 ft (305 m) in 50 sec
Service ceiling: 20,500 ft (6250 m)

The Triplane was produced immediately after the Pup and was used almost exclusively by the Royal Naval Air Service. The first Triplane-equipped unit, No.1 (Naval) Squadron, went into action in February 1917. Five RNAS units used the type in France and one machine was used in the Aegean theatre from Mudros.

The Triplane's design was a complete departure from all previous military aircraft and it was adopted to give a high degree of manoeuvrability as well as the best possible field of vision to its pilot. It had fine flying qualities (see "Flying the Museum's Aircraft" by W/C Hartman for further flight comments) and was a most effective fighter, so much so that on its appearance German pilots immediately asked for triplane designs, which resulted in a rash of German triplanes. Only one of them, the Fokker Dr.I, went into limited production. It is ironic that the Fokker triplane is well known today because Germany's leading fighter pilot, Baron von Richthofen, was killed in one, while its inspiration, and a better flying machine, the Sopwith Triplane, remains much in the background.

The Triplane was superseded by the more heavily armed Sopwith Camel about the beginning of September 1917. Many Canadians flew the Triplane but Raymond Collishaw was its leading exponent. He led the all-Canadian B Flight of No.10 (Naval) Squadron, known as the Black Flight because of its colour marking and aircraft names. Members of the Black Flight and their aircraft names were F/S/L R. Collishaw of Nanaimo, British Columbia, *Black Maria*; F/S/L W.M. Alexander of Toronto, *Black Prince*; F/S/L G.E. Nash of Stoney Creek,

Ontario, *Black Sheep*; F/S/L E.V. Reid of Toronto, *Black Roger*; and F/S/L J.E. Sharman of Oak Lake, Manitoba, *Black Death*. Reid and Sharman were killed and Nash was made a prisoner of war during the summer of 1917. This outstanding team has been called "as redoubtable a fighting unit as took to the air during the war" and as accounting "for no fewer than 87 enemy aircraft between May and July 1917." This sterling performance has made it difficult for any reasonable history of the air war to omit at least a reference to the Black Flight of No.10 (Naval) Squadron and its Sopwith Triplanes.

The Museum's specimen is a reproduction built by C.R. Swanson of Sycamore, Illinois, during 1963-66. Its engine was supplied by the Museum and it was first flown on May 5, 1967, by W/C P.A. Hartman at Rockcliffe. It was flown occasionally until 1971.

Sopwith *2F.1 Camel*

SPECIFICATIONS

Powerplant: One 130 hp Clerget 9B
Wing span, upper and lower: 26 ft 11 in (8.2 m)
Length: 18 ft 6 in (5.6 m)
Height: 9 ft 1 in (2.8 m)
Wing area: 221 sq ft (20.5 m²)
Weight, empty: 956 lb (434 kg)
Weight, gross: 1,523 lb (691 kg)
Speed, max.: 114 mph at 10,000 ft (184 km/h at 3050 m)
Climb: 6,500 ft (1980 m) in 6 min 25 sec
Service ceiling: 19,000 ft (5790 m)

The Camel was the most numerous of all British World War I fighters and accounted for more enemy aircraft than any other type. It was made in two forms, the F.1 Camel for the RFC and the 2F.1 for the

RNAS. The 2F.1 differed from the F.1 in having a wing span 13 inches shorter, a fuselage made in two pieces that could be quickly separated for shipboard use, and an over-wing Lewis gun and a single Vickers gun on the fuselage in place of the two fuselage-mounted guns of the F.1. The type's name was derived from the hump that enclosed the breeches of the Vickers guns.

With the weight of the engine, guns, fuel and pilot located in very close proximity, the Camel was highly manoeuvrable. It was unstable and proved tricky, sometimes lethal, in the hands of inexperienced pilots. W/C P.A. Hartman describes flying the Camel as "like balancing an egg on the point of a needle." However, experienced pilots liked its manoeuvrability and turned its idiosyncrasies to advantage in dogfights. It did not endear itself to all pilots and has been called a "popping little firecracker" and worse by its detractors, usually S.E.5A pilots who did not care for the "fierce little beast."

The F.1 Camel attained fame in France and all other war theatres. Many RFC/RAF aces flew Camels, and the leading Canadian exponent of the type was Major W.G. Barker of Nos. 28 and 66 Squadrons, who destroyed 41 enemy aircraft in it. All these were accounted for in a single Camel, B65313, a feat without equal which speaks eloquently for his flying skill. Major D.R. MacLaren of 46 Squadron was another Canadian Camel ace, with 54 enemy aircraft to his credit, all on Camels.

The 2F.1 Camels were flown largely from carriers and some from towed lighters. These operations rarely made headlines, although Lt. S.D. Culley from South Africa (frequently and erroneously noted as a Canadian) shot down Zeppelin L.53 while flying a 2F.1 Camel from a lighter.

The Museum's specimen was built by Hooper and Co. of London in 1918 and was one of the last Camels made. It served in the RAF until coming to Canada in 1925 along with six others. These were not flown in Canada but were used for ground training. The Museum's aircraft was donated to the Aeronautical Museum and displayed from 1936 to 1940, then stored. It was loaned to the Canadian War Museum in 1957 and was restored unsatisfactorily and displayed at various places in 1959. It was restored to flying condition in 1966-67 by C.R. Swanson, Sycamore, Illinois, and finished in markings of a Camel flown by Lt. W.S. Lockhart from HMAS *Sydney* in early 1919. It was first flown at Rockcliffe, May 26, 1967, by W/C Hartman and has been displayed since then at the Canadian War Museum.

Sopwith 7F.1 Snipe

SPECIFICATIONS

Powerplant: One 230 hp
             Bentley B.R.2
Wing span: 31 ft 1 in (9.5 m)
Length: 19 ft 10 in (6 m)
Height: 9 ft 6 in (2.9 m)
Wing area: 271 sq ft (25.2 m²)
Weight, empty: 1,312 lb (595 kg)
Weight, gross: 2,020 lb (916 kg)
Speed, max.: 121 mph at 10,000 ft
             (195 km/h at 3050 m)
Climb: 6,500 ft (1980 m)
         in 5 min 10 sec
Service ceiling: 19,500 ft (5940 m)

The Snipe was the last British fighter to enter squadron service in World War I. It was intended as a replacement for the Camel but only three squadrons were equipped with it by the Armistice.

The type had a number of development problems and several different prototypes were made with various modifications until it was ordered into production in March 1918. However, design changes were still being made after production started. The type, in spite of its increased power, offered very little improvement over the Camel. The new enemy fighters would have been considerably better and, had the war continued, it would have been quickly replaced by the fine Martinsyde F.4 Buzzard, but the Snipe soldiered on in the postwar RAF until as late as 1927 as it was readily available from the war production.

The Snipe has a firm place in the history of the first war in the air owing to a single combat on October 27, 1918, when Major W.G. Barker of Dauphin, Manitoba, attached to No.201 Squadron with his Snipe, encountered several groups of enemy aircraft. The battle that followed and that of Lt. Werner Voss with No.56 Squadron on September 23, 1917, are regarded as the two classic single-handed air battles of the war. Barker's fight was a drawn-out affair. After downing a two-seater, he was wounded in the thigh by a Fokker D.VII, then spun into an enemy group of 15 machines. He attacked two inconclusively and shot a third down in flames. Again wounded, he fainted and revived in the midst of another 12 or 15 enemy machines. He got another in flames, but his left elbow was shattered by a bullet in the encounter. Fainting again, he revived and attempted to ram a Fokker but then shot it down, again in flames. Attempting to escape, he was surrounded by eight enemy machines, which he eluded, then crashed after regaining the British lines. He was awarded the Victoria Cross for this action and seldom has this medal been better deserved.

The Museum's specimen was built by Nieuport and General Aircraft Ltd., Cricklewood, England. Details of its RAF history are lacking but it likely served abroad as did most RAF Snipes. It was brought to the United States in 1926. For details of its North American career see Appendix 2. It suffered minor damage during a flying display but was exhibited until 1987, when it was restored and refinished.

Spad 7

SPECIFICATIONS

Powerplant: One 180 hp
            Hispano-Suiza 8Ab
Wing span, upper: 7.82 m
            (25 ft 7 7/8 in)
Wing span, lower: 7.57 m (24 ft 10 in)
Length: 6.1 m (20 ft)
Height: 2.3 m (7 ft 6-1/2 in)
Wing area: 17.85 m² (192 sq ft)
Weight, empty: 500 kg (1,102 lb)
Weight, gross: 705 kg (1,554 lb)
Speed, max.: 193 km/h at 2000 m
            (120 mph at 6,560 ft)
Climb: 2000 m (6,560 ft) in 4.66 min
Service ceiling: 6553 m (21,500 ft)

The Spad 7 was designed by the Société anonyme pour l'Aviation et ses Dérivés at Paris to exploit the excellent Hispano-Suiza engine and was the first fighter to use it. The prototype first flew in April 1916 and the type was adopted as the standard French fighter and built at eight French factories. All the renowned French fighter pilots flew Spads, and later it was the standard fighter of the US Signal Corps in France. In addition it was adopted by the Belgian, British, Italian and Russian air forces and built under licence in Britain and Russia. As one might expect with this history, the Spad 7 has become one of the classic fighters of World War I.

For unknown reasons, the Spad never distinguished itself in the RFC in the way that might be expected from its success in other air services. It was used by Nos. 19, 23 and 60 Squadrons in France, Nos. 30 and 63 Squadrons in Mesopotamia, and No.72 in Palestine, as well as in a number of training squadrons. The highest scoring RFC Spad pilot was Major A.D. Carter of Pointe de Bute, New Brunswick, who scored 12 of his 22 victories on Spads. (This number is believed to be correct. Both 27 and 31 have also been reported but cannot be substantiated.)

The Museum's specimen was built by Mann, Egerton and Co. Ltd., Norwich, England, and apparently served with training squadrons before being brought to the United States. It was at Rockwell Field, San Diego, in 1919. Col. J.B. Jarrett, who operated a museum on the Steel Pier at Atlantic City, New Jersey, obtained the Spad from an aviation junk dealer, Arrigo Balboni, in California, who stated that it had been used in the motion picture *Wings*. Jarrett patched it up for display and later got the remains of another Spad and used them to improve the first. During World War II Col. Jarrett's aircraft collection was stored under poor conditions and deteriorated. In 1949 it was sold to Frank Tallman in California. In the 1950s Tallman gave the Spad to James Petty of Gastonia, North Carolina, in return for Petty's restoration of a Phalz D.XII. Petty then restored the Spad and it seems likely that few of the original wooden parts now remain.

The aircraft was loaned to the USAF Museum for display and is believed to have been flown there by Major Kimbrough S. Brown with a borrowed radiator. It was offered for sale and in December 1965 the National Aviation Museum was fortunate in obtaining this example of a classic aircraft. It is still finished in the markings of the distinguished Escadrille SPA.3 of Aviation militaire that it bore when it was obtained. It is likely that when refinished it will be done as a Mann, Egerton machine of the RFC as it was originally (see page 46).

On September 21, 1929, the prototype Stearman Model 4 was first flown at Wichita, Kansas–about a month before the stock market crash that brought on the Depression. It was also just before American airmail policy began to change to encourage the carrying of passengers rather than mail alone, and not long before the introduction of the modern low-wing monoplane. So it was hardly an auspicious time to launch a new aircraft, even though Lloyd Stearman claimed it was the best aircraft he ever designed and others feel it marked the ultimate development of the civil biplane in North America.

The Model 4 was made in three basic versions powered by the

Stearman 4EM Junior Speedmail

SPECIFICATIONS

Powerplant: One 450 hp Pratt &
Whitney Wasp SC
Wing span, upper: 38 ft (11.6 m)
Wing span, lower: 28 ft (8.5 m)
Length: 26 ft 4 in (8 m)
Height: 10 ft 2 in (3.1 m)
Weight, empty: 2,455 lb (1113 kg)
Weight, gross: 3,936 lb (1785 kg)
Speed, max.: 158 mph (254 km/h)
Speed, cruising: 134 mph (216 km/h)
Climb: 1,400 ft/min (427 m/min)
Ceiling: 18,000 ft (5490 m)

300 hp Wright Whirlwind or 300 hp Pratt & Whitney Wasp Jr. or 420/ 450 hp Pratt & Whitney Wasp C/SC, and each could be supplied as a three-seat open-cockpit machine or as an open-cockpit mailplane to carry about 600 lb (270 kg) of mail. The Model 4 was the first aircraft to enter production with the new NACA engine cowl and its elegant lines were much admired. Forty were built, with several going to oil companies for fast commuting between their head offices and the oil fields, a few to wealthy private owners and the balance to mail carriers. The largest fleet was operated by American Airlines. In their later lives many were converted for use as crop dusters and soon established themselves as real workhorses, although they certainly lost their original elegant appearance in this role.

Canadian Airways operated four 4EM mailplanes on their eastern Canada routes in the early 1930s. They were the fastest aircraft in Canadian skies, even edging out the Siskin fighters of the RCAF, and when introduced on new routes invariably set new records. Their stability and fine flying characteristics made them great favourites with the pilots. Two Whirlwind-powered 4Cs were used by the Civil Aviation Branch of the Department of National Defence for airway inspection purposes.

The Museum's specimen was one of three 4Es supplied to the Standard Oil Co. of California in June 1930. It was converted to a crop duster about 1940 by Carberry Crop Dusters Inc. of Fresno, California, and used by them until acquired by Dean Wilson of Homedale, Idaho, from whom it was purchased by John Paterson of Fort William on January 25, 1965. Paterson first flew the restored aircraft in December

1969. It was flown to Rockcliffe with some commemorative mail and presented to the Museum on September 25, 1970. It has been on display ever since as a fine reminder of those long-gone days when pilots donned helmets and goggles and parachutes and flew the mail without the benefit of radio and other modern amenities.

Stinson SR Reliant

SPECIFICATIONS

Powerplant: One 215 hp
           Lycoming R-680*
Wing span: 43 ft 3 in (13.2 m)
Length: 27 ft (8.2 m)
Height: 8 ft 5 in (2.6 m)
Wing area: 235 sq ft (21.8 m²)
Weight, empty: 2,070 lb (939 kg)
Weight, gross: 3,155 lb (1431 kg)
Speed, max.: 130 mph (209 km/h)
Speed, cruising: 115 mph (185 km/h)
Climb: 750 ft/min (229 m/min)
Service ceiling: 14,000 ft (4270 m)

Edward A. Stinson learned to fly in 1915, encouraged by his sisters, Katherine and Marjorie, who were both pilots. In 1925 he formed a syndicate and built a cabin biplane at Detroit. It had an enclosed and heated cabin, self starter and wheel brakes, and while none of these features was new it was the first time that all were incorporated in one aircraft. The aircraft attracted favourable attention and a company, the Stinson Aircraft Corp. was formed in May 1926.

The new company marketed a production version of his biplane, designated SB-1 Detroiter. A number were sold, including two to Hubert Wilkins (later Sir Hubert) for use in exploration. However, it was just then that the new, more efficient monoplanes were being introduced, so Stinson dropped the biplane and introduced his SM-1 Detroiter monoplane in 1927, the first of a long line of Stinson monoplanes.

The Stinson SR Reliant first flew in 1933 and was intended to suit

* Original equipment; it now has 300 hp
Lycoming R-680 E-3 fitted.

private owners and small charter operators. Eighty-nine were sold in the difficult Depression years, plus seven special versions given SR numbers -1 to -4. These became the first of a series of aircraft also marketed under the name Reliant although most were quite different from the original SR.

The Stinson SR never became popular in Canada, but why this was so is unknown. Certainly there was a market for aircraft in its class, as the four-place Waco cabin biplanes were widely used here in the 1930s. Initially only one SR came to Canada and it was the prototype, which became CF-AUS. It was imported by K. Johanesson Flying Service of Winnipeg and served in Manitoba with Johanesson and T. Lamb until 1937. It then flew in Saskatchewan until withdrawn from use in 1943.

The Museum's specimen was made in 1933, the 17th SR built. It was registered NC13464 and passed through the hands of a number of American owners until bought by I.I. Handberg in 1953 and registered CF-HAW to Lac la Croix-Quetico Air Service of Lac la Croix, Ontario, the second and likely the last SR registered here. It then passed through three other owners before being restored in 1964 to suit Ellis Culliton of Bracebridge, Ontario. The cabin was enlarged and lined with aluminum sheet, new cabin doors were installed and a pilot's door was added. The original plate-glass windows were replaced by Plexiglas ones and a new full instrument panel was added. The aircraft was then sold to Akela Aircraft Repair, Whitefish, Ontario, from which it was bought by the Museum and flown on floats to Rockcliffe on June 1, 1983.

Stits SA-3A *Playboy*

The Stits Playboy is one of numerous small aircraft that are today generally classified as "homebuilts." Lest one should think that "homebuilt" implies these machines are roughly made with little regard for fine workmanship and good aircraft practice, it should be said that this is far from the case and most exhibit excellent, and frequently superb, workmanship.

In the pioneer days most aircraft in North America were homebuilt, as there were few aircraft companies; the Museum's McDowall Monoplane is a pioneer homebuilt. In the 20s and 30s many homebuilts were made from kits or parts and others from basic materials. Popular designs included the Corben Baby Ace, Heath Parasol Monoplane and Pietenpol Air Camper. Most were powered with converted automobile or motorcycle engines.

Making homebuilts virtually ceased for a short period following

SPECIFICATIONS

Powerplant: One 100 hp
          Lycoming 0-235-C1
Wing span: 22 ft (6.7 m)
Length: 17 ft 4 in (5.3 m)
Height: 4 ft 9 in (1.4 m)
Wing area: 96 sq ft (8.9 m²)
Weight, empty: 685 lb (311 kg)
Weight, gross: 960 lb (435 kg)
Speed, max.: 150 mph (241 km/h)
Speed, cruising: 130 mph (209 km/h)
Climb: 1,000 ft/min (305 m/min)
Service ceiling: 12,000 ft (3660 m)

World War II, and started again in the United States in the 1950s, rapidly becoming popular, and today the building and flying of home-builts is a sizable portion of total aviation activity. The Experimental Aircraft Association was formed to further this activity and provides assistance in many ways to its members to ensure that they produce the best and safest product possible. It soon had chapters in the United States, Canada and other countries. Soon the Experimental Aircraft Association of Canada (originally Ultra Light Aircraft Association of Canada) was formed to serve specific Canadian interests in this field.

The first Canadian to undertake the construction of a homebuilt aircraft in the postwar period was the late Keith S. Hopkinson of Goderich, Ontario. He selected the Stits Playboy for his project in 1954. He introduced some modifications to the design, incorporating a nose cowl from a Piper J-3, a spinner from a Cessna 170, wing struts from a D.H.82C Tiger Moth, landing gear from a Cessna 140, and Stinson 108 wheel pants. Modifications of this sort were typical of the homebuilder.

The aircraft took only 11 months to complete and is believed to have first flown in September 1955. It was known to "Hoppy" as *Little Hokey.* As a result of his dealings with the Department of Transport over the building of *Little Hokey,* Hoppy is credited with persuading the DOT to facilitate the licensing of homebuilts in Canada. After Hoppy's death his Playboy was sold to D.A. Kernohan, and in 1978 it was bought by the Museum as the first and only modern example of this popular type of aviation in the collection.

Supermarine Spitfire

The **Supermarine Spitfire** IX L.F. was introduced in 1942 to counter the latest German fighters that were outperforming the Spitfire V. The main change was the addition of a Merlin with a two-stage supercharger which improved the high-altitude performance. There were two main variants: the IX L.F. for low-altitude duties with the wingtips removed, and the IX H.F. for high-altitude duties with the wingtips installed.

(DND PCN 5234)

The **Supermarine Spitfire** II was a development of the Mk. I which appeared in 1940, but only a few flew in the Battle of Britain. The main change was the slightly more powerful Merlin III replacing the Merlin II of most Spitfire Is. The IA had four 303 Browning guns and the IIB had two 20 mm cannons in addition.

(DND 1476-2)

The **Supermarine Spitfire** XVIe appeared late in 1944 and differed from the Mk. IX in having the Packard Merlin 266 fitted to replace the Rolls-Royce Merlin 66. Like the IX it was made in two versions: a low-altitude version, the XVI L.F. without wingtips and the high-altitude version XVI with wingtips. Some XVIs had a rounded rudder like the IX and others, like the Museum specimen, had a pointed rudder.

(DND, RNC 1538-46)

The design of the Spitfire was started in November 1934 as a private venture by Supermarine Aviation Works (Vickers) Ltd. but the following month the company received a contract for a prototype. The prototype first flew on March 6, 1936. The type was ordered into production in June, and the first production aircraft flew on May 15, 1938.

The Spitfire became the most famous British fighter of World War II and was a pilot's delight to fly. It was developed through 24 versions to suit specific duties and to enhance its performance to compete with German fighter development, and was in production from 1938 to 1947. Much of the improvement in performance was due to the ever-increasing power of new versions of the Rolls-Royce Merlin engine and, finally, the even more powerful Rolls-Royce Griffon. No other World War II fighter had such a long production life; in all, 20,334 Spitfires and 1,220 Seafires, the naval version, were made.

The Spitfire first saw action against the Luftwaffe in October 1939 but was rarely in action until the Battle of Britain began in July 1940. Then 18 Spitfire squadrons joined 26 Hurricane, five Blenheim and two Defiant squadrons to counter the German attack. The Spitfires would normally deal with the higher-flying German fighters while the Hurricanes took care of the lethal bombers at lower altitudes. Their pilots were the "few" to whom so much is owed. Following the Battle of Britain, the numbers of Spitfires built up and the type became the standard British single-engined high-altitude fighter, with the Hurricane being used almost entirely at low altitudes.

The Museum is fortunate in having three different Spitfires, a Mk.IIB, a Mk.IX L.F. and a Mk.XVIe, all made at the Castle Bromwich Aircraft Factory, the largest Spitfire producer. Both the Mk.IIB and Mk.IX L.F. have interesting histories. The Mk.IIB, presented to the RAF during the war by the State of Soebang, Netherlands East Indies, was used by 222 Squadron RAF in 1941 in sweeps over northern France until its Mk.IIBs were replaced by Mk.VBs. It came to Canada in 1942 and went to Mountain View, Ontario, and Camp Borden, then to Station Uplands in 1962 and the RCAF historic aircraft collection in 1964. It is now in the markings of an aircraft of 45 Squadron RAF.

The Mk.IX L.F. served with No.308 (Polish) and 416 RAF squadrons in 1944. After the war it was sold to the Netherlands Army Air Force and sent to the East Indies. Then it was sold to the Belgian Air Force and was registered civilly there before being bought by J.N. Paterson of Fort William, Ontario. It was registered civilly and flown by Paterson, who gave it to the Museum in 1964.

The Mk.XVIe was built in April 1945 and never saw action.

SPECIFICATIONS (Mk.IX L.F.)

Powerplant: One 1,250 hp Rolls-Royce Merlin 76*

Wing span: 32 ft 7 in (9.9 m)

Length: 31 ft 4 in (9.5 m)

Height: 12 ft 7-1/4 in (3.8 m)

Wing area: 231 sq ft (21.5 m²)

Weight, empty: 5,800 lb (2630 kg)

Weight, gross: 7,500 lb (3402 kg)

Speed, max.: 404 mph at 21,000 ft (650 km/h at 6400 m)

Speed, max. cruising: 328 mph at 20,000 ft (528 km/h at 6100 m)

Climb: 20,000 ft (6100 m) in 6.4 min

Service ceiling: 42,500 ft (12 950 m)

* Rolls-Royce Merlin 66 originally fitted to Mk.IX L.F.

Taylor E-2 Cub

SPECIFICATIONS

Powerplant: One 40 hp
Continental A-40-4
Wing span: 35 ft 3 in (10.7 m)
Length: 22 ft 3 in (6.8 m)
Height: 6 ft 6 in (2 m)
Wing area: 184 sq ft (17.1 m²)
Weight, empty: 525 lb (238 kg)
Weight, gross: 925 lb (420 kg)
Speed, max.: 78 mph (126 km/h)
Speed, cruising: 65 mph (105 km/h)
Climb: 450 ft/min (137 m/min)
Ceiling: 12,000 ft (3660 m)

The Taylor Brothers Aircraft Corp. was originally formed at Rochester, New York, in 1926 by C.G. and Gordon Taylor and made a light, side-by-side, parasol monoplane named the Chummy to a design of C.G. Taylor. The company moved to Bradford, Pennsylvania, in 1929 after the death of Gordon. The success of the Aeronca C-2 prompted the company to introduce a new design, a tandem two-seat light aircraft arbitrarily designated the E-2 and named the Cub. The company became the Taylor Aircraft Co. in 1931 and the Piper Aircraft Co. in 1937.

The E-2 prototype first flew in September 1930. It was licensed in 1931 and 24 were sold that year at a price of US$1,325 in spite of the Depression. Production of the E-2 stopped in early 1936 with just under 350 made. Its role was private and instructional flying and it proved popular, but the importance of the E-2 Cub was that it led to the development of the Taylor J-2 Cub, an improved version, late in 1935 and to the Piper J-3 Cub of 1937, a still further improvement. Both of these became very popular and made the name Cub almost a synonym for a light aircraft in the minds of many people.

As only three E-2 Cubs were imported into Canada before the war, the type had little influence on Canadian aviation. However, Cub Aircraft Ltd. was formed at Hamilton, Ontario, in 1936 to sell and

208

service J-2 and J-3 Cubs, and J-3s appeared in this country in good numbers. Following the war, J-3s were made at Hamilton and many J-3 and a few J-2 Cubs remain active in this country. Two E-2 Cubs were imported in the postwar period, including the Museum's.

The Museum's specimen was built by the Taylor Aircraft Co. in August 1935 and registered to its first owner, Gordon Curtis of New York City, on September 6. It had four more owners before going in 1940 to the Choate School, a boys preparatory school at Wallingford, Connecticut, where it was used for ground instruction. Garth Elliot of Toronto bought it in 1975 and meticulously restored it with the assistance of Max Say and the co-operation of Frank Shaineline and D.J. (Dick) Berg of the Department of Transport. During this time the wings were destroyed by fire and a new set was obtained in Ohio. The Cub was first flown in Canada on July 30, 1980. The Museum obtained the aircraft in June 1985 in exchange for its unrestored example, so that it would be available for display in the new building.

Travel Air 2000

SPECIFICATIONS

Powerplant: One 90 hp Curtiss OX-5 (100 hp Curtiss OXX-6)

Wing span, upper: 34 ft 8 in (10.6 m)
Wing span, lower: 28 ft 8 in (8.7 m)
Length: 24 ft 2 in (7.4 m)
Height: 8 ft 11 in (2.7 m)
Wing area: 297 sq ft (27.6 m²)
Weight, empty: 1,335 lb (605 kg)
Weight, gross: 2,180 lb (989 kg)
Speed, max.: 100 mph (161 km/h)
Speed, cruising: 85 mph (137 km/h)
Climb: 520 ft/min (158 m/min)
Ceiling: 10,000 ft (3050 m)

In the 1920s the three-place open-cockpit biplane powered with a Curtiss OX-5 engine dominated American general civil aviation. Many similar designs were on the market, the most numerous being the Alexander Eaglerock, Travel Air 2000 and the Waco 9 and 10. They were used for flying instruction, barnstorming, private flying, charter trips and similar activities.

The Travel Air 2000 was built by the Travel Air Manufacturing Co. of Wichita, Kansas. The prototype, the Model 1000, first flew in 1925 and with minor improvements went into production as the Model 2000. The prototype won first prize for efficiency in the First National (Ford) Air Tour of 1925, and this and its subsequent acceptance are not surprising when it is realized that Walter Beech, Clyde Cessna and Lloyd Stearman were all involved in its design and manufacture and each went on to found a successful aircraft company bearing his own name.

The 2000 was widely and affectionately known as "Old Elephant Ears" owing to its balanced ailerons and had perhaps the most distinctive appearance in its class. The type was often used by Hollywood in the late 1920s and early 1930s as a stand-in for the famous World War I fighter, the Fokker D.VII, which it resembled. Because of this the type was frequently referred to as the "Wichita Fokker" in deference to its place of origin.

In Canada, the Waco 9 and 10 and the Travel Air 2000 were the most popular of their class. The Continental Aero Corporation at St. Hubert Airport, near Montreal, obtained the Travel Air sales rights for Canada in 1928 and imported various Travel Air types before going out of business in the Depression. They operated a flying school with 2000s at St. Hubert, and among the 2000s sold was one to the Walter M. Lowney Co., a maker of chocolates in Montreal – an early example of a Canadian business aircraft.

The Museum's specimen, CF-AFG, was one of two 2000s bought in September 1929 by Janney Aircraft and Boats Ltd. of Kingston, Ontario. It passed through several hands until it was stored in 1941. It was sold in 1958 to Keith Hopkinson of Guelph, Ontario, and was purchased by the Museum in September 1968 from his estate. It is complete but needs general restoration and re-covering prior to display. A curiosity is that at some time a 100 hp Curtiss OXX-6 engine was substituted for its original OX-5 without, apparently, being reported to the Department of Transport.

Vickers 757 Viscount

The Viscount originated from a proposal by Vickers-Armstrong Ltd. of Weybridge in the spring of 1945. Contracts were issued by the British Ministry of Supply for its development and the first prototype, the Type 630, first flew in July 1948. Following the construction of two more prototypes, the first production version first flew in August 1950. This version was considerably larger than the original both in dimensions and weight. The Viscount then became the first propeller turbine

SPECIFICATIONS

Powerplant: Four 1,400 hp
         Rolls-Royce Dart 506

Wing span: 93 ft 8-1/2 in (28.6 m)

Length: 81 ft 9 in (24.9 m)

Height: 27 ft 9 in (8.5 m)

Wing area: 963 sq ft (89.5 m²)

Weight, empty: 38,061 lb (17 264 kg)

Weight, gross: 63,000 lb (28 577 kg)

Speed, max.: 380 mph at 20,000 ft
         (611 km/h at 6100 m)

Speed, cruising: 324 mph at 20,000 ft
         (521 km/h at 6100 m)

Climb: 1,200 ft/min (366 m/min)

Ceiling: 28,500 ft (8690 m) with
         rate of climb 200 ft/min
         (61 m/min)

aircraft to enter production and in April 1953 it began the first propeller turbine airline service, over British European Airways' London-Rome-Athens-Nicosia route.

Trans-Canada Air Lines became interested in the type but requested a long list of changes to make it suitable for North American airline operations. The changes were agreed to and an original TCA order for 15 Type 724 Viscounts was followed by orders for 36 more 757 Viscounts for a total of 51. The TCA modifications made the type acceptable to both Capital and Continental Airlines in the United States, and the Viscount became the first British commercial transport to enter American service.

TCA took delivery of its first Viscount in December 1954 and in April 1955 inaugurated regular airline service with the type, the first propeller turbine aircraft service in North America. The Viscount proved very popular with its increased speed, lack of noise and vibration, and large windows providing an unusually good view for the passengers, especially in the 40-seat version first bought by TCA. Viscounts were used on most domestic routes and remained in service with the airline until 1969. The Viscount provided reliable service with few maintenance problems, and the ever-increasing life between overhauls of the Rolls-Royce Dart engines made the type economical to operate.

The Museum's specimen was delivered to TCA in May 1957 and continued to serve on their routes until retired in 1969. It was presented to the Museum on November 19, 1969, by Air Canada, refinished in its original TCA colours. It has had to be stored outside since then owing to lack of space.

Westland Lysander III

SPECIFICATIONS

Powerplant: One 870 hp Bristol
           Mercury XX
Wing span: 50 ft (15.2 m)
Length: 30 ft 6 in (9.3 m)
Height: 11 ft 6 in (3.5 m)
Wing area: 260 sq ft (24.2 m²)
Weight, empty: 4,670 lb (2118 kg)
Weight, gross: 6,100 lb (2766 kg)
Speed, max.: 217 mph at 10,000 ft
           (349 km/h at 3050 m)
Speed, cruising: 170 mph at 15,000 ft
           (274 km/h at 4570 m)
Climb: 20,000 ft (6100 m)
      in 21.7 min
Service ceiling: 23,850 ft (7270 m)

The Lysander was designed to meet a 1934 RAF specification for an army co-operation aircraft. It was envisaged that this aircraft would be able to operate from short, unprepared fields close to the troops, with what is today termed good STOL performance. To meet this requirement Westland Aircraft Ltd. of Yeovil, England, designed a high-wing monoplane with full-span wing slots which automatically controlled the flaps. The prototype first flew in June 1936 and the aircraft was put into production and entered RAF service in 1938.

The RCAF selected the Lysander for its army co-operation requirements and placed an order with the Aircraft Division of National Steel Car Corp. at Malton, Ontario. The Canadian prototype first flew on August 16, 1939. With the outbreak of war, large additional Canadian orders were planned both for the RAF and India. However, when the "phoney war" ended in the spring of 1940, it was soon found that the type was unsuited to modern warfare and the RAF's Lysanders were diverted to air-sea rescue operations and other duties, and to the subsequently much publicized role of delivering and picking up secret agents in occupied countries, for which they were well suited.

The RCAF's Lysanders received much publicity while being used by No.110 Squadron working up to go overseas, and a few were used briefly on the east and west coasts on maritime patrol duties. Shortly, all RCAF Lysanders were relegated to target towing duties with the British Commonwealth Air Training Plan and they served at all bombing and gunnery schools. Many were sold as surplus at the end of the war and four were used for crop spraying, the only Lysanders to enter civil use. A number of Canadian Lysanders survived and can be seen in several of the world's aviation museums.

The Museum has had two Lysanders, the first having been sent to India, where it is now in their museum at New Delhi, as related in the history section of this book. The present Lysander was assembled and restored from British- and Canadian-built components by Capt. B. Lapointe and others at Winnipeg and was first flown by him on December 29, 1967. It was flown at several locations in western and central Canada before being flown to Rockcliffe and presented to the Museum in September 1968. It is finished in the markings of a British Lysander I serving with 110 Squadron RCAF in England.

Zenair CH300 Tri-Zenith

SPECIFICATIONS

Powerplant: One 180 hp
Lycoming 0-360-AFZ
Wing span: 26 ft 6 in (8.1 m)
Length: 22 ft 6 in (6.9 m)
Height: 6 ft 10 in (2.1 m)
Wing area: 130 sq ft (12.1 m²)
Weight, empty: 1,140 lb (517 kg)
Weight, gross: 1,850 lb (839 kg)
Speed, max.: 170 mph (274 km/h)
Speed, cruising: 153 mph (246 km/h)
Climb: 1,400 ft/min (427 m/min)

Christophe Heintz came to Canada in 1973 after serving as a design engineer in France. He shortly set up his own company, Zenair Ltd., in partnership with G. Boudreau. The object was to design and build light aircraft; now they produce ultralight aircraft, suitable for the home builder and sold both as kits of material and as partially completed parts. The company has developed and marketed eight designs. About 250 aircraft have been completed and about 500 more are under construction in 34 different countries.

Zenair aircraft have a number of awards and accomplishments to their credit, including: 1974, best new design awarded by Experimental Aircraft Association of Canada for the CH200; 1975, winner of Pazmany two-seat efficiency contest; 1975, National Association of Sport Aircraft Designers seal of quality awarded for their drawings and kits; in 1976, a CH200 was assembled and flown in eight days at Oshkosh, Wisconsin; in 1978, the Museum's CH300 set a new official record for its class by flying Vancouver-Halifax non-stop.

The CH300 prototype was first flown in 1977 and became available in 125, 150 and 180 hp models. It has a standard tricycle undercarriage with an optional three-point gear available. It is capable of carrying three people but current Canadian regulations limit it to two people with a very generous 210 lb (95 kg) baggage allowance.

The Museum's specimen was built by Robin (Red) Morris for his trans-Canada flight with the assistance of G. Boudreau and D. Holtby, and Red first flew it on May 27, 1978. It was a standard CH300 with the following changes. The normal fuel capacity of 28 Imperial gallons (127 L) was increased to 170 gallons (773 L). A specially designed seat was fitted for pilot comfort. IFR instrumentation was fitted together with an Edo-Aire single-axis autopilot. These changes meant the aircraft would take off at 2,710 lb (1230 kg) gross weight, or 860 lb (390 kg) over its normal maximum weight. This and the fact it would fly under instrument flight rules made it necessary to obtain special waivers from the Department of Transport. Financial problems were overcome with the assistance of the Pepsi-Cola Co., Edo-Aire Ltd. and Zenair Ltd.

Red Morris left Vancouver at 11:30 a.m., July 1, 1978, and landed 22 hours, 43 minutes and 43 seconds later at Halifax. The flight was uneventful, although headwinds near the Manitoba-Ontario border increased fuel consumption and the alternator failed, which required careful handling of the remaining battery energy. A distance of approximately 2,800 miles (4500 km) was flown at about 130 mph (210 km/h), a new record for this class of aircraft. It was the first officially observed Vancouver-Halifax flight. The aircraft was sold to the Museum and was flown there by Red Morris on October 25, 1984.

Brief descriptions follow of a few major components and unrestored machines.

Airspeed Consul

The Consul was a British civil conversion of the Oxford World War II twin-engined trainer. It was acquired by the Canadian War Museum in 1965 and brought to Rockcliffe the following year with the intention of reconverting it to its Oxford trainer configuration. This has not been done and it has never been assembled or exhibited. It is now considered unlikely that it will be reconverted and exhibited.

Blériot XI

This aircraft was built in San Francisco, California, in 1911 for J.W. Hamilton and first flown that July. It is believed to be the first successful California-built heavier-than-air machine and is powered by a four-cylinder Elbridge water-cooled engine. It was purchased in December 1971 in California by the Museum and is complete except for wheels and radiator. It is likely that it will be restored at an early date but there are at present no plans for this.

Bristol Beaufighter T.F.X.

An airframe of this notable British World War II fighter was received from the RAF in exchange for a Bristol Bolingbroke in 1969. However, it is without engines and nacelles, which will have to be found before it can be restored and displayed. Prospects for this happening do not seem good.

De Havilland D.H.82C Tiger Moth

The Museum holds most airframe components of Tiger Moth CF-FGL, construction number 724. It is unlikely that they will ever be built into a complete aircraft owing to the similarity to the Menasco Moth in the collection. Probably they will form part of an exhibit on aircraft structures in the future; in fact, one wing was restored for this purpose in 1966-67 but the exhibit was never completed.

De Havilland DHC-6 Twin Otter

The Museum has held the airframe of the prototype less engines since it was donated by de Havilland Canada in 1981. Engines have now become available, enabling it to be completed and displayed.

Fleet 2/7

The Museum holds many components of a Fleet trainer, which could be a Fleet 2 or 7 depending on the powerplant installed. It is unlikely that they will be made into a complete aircraft owing to the similarity to the Fleet 16B Finch IIB already on display. It is likely the components will be used in an aircraft structures display in the future.

The Museum's prototype of the de Havilland DHC Twin Otter on its initial flight on May 20, 1965.
(DHC 21040)

Fleet 50 Freighter

A damaged airframe was salvaged in northern Quebec by the Museum in 1964 and an additional wing and small components were salvaged in Ontario in 1968 by M.L. McIntyre. Restoration has not yet been attempted and it is unlikely it will be started in the foreseeable future.

Sikorsky R-4B

As mentioned elsewhere, the airframe of this helicopter, the world's first production helicopter type, was obtained in an exchange with an American museum in 1983. It is without a powerplant but it is believed one can be obtained, and, of course, it needs restoration.

Wills Wing XC-185 Hang Glider

This is the Museum's only example of the hang glider, which has proved so popular in recent years as a recreational vehicle. It was donated by its owner, Stewart Midwinter of Calgary, in 1982. It has not been displayed.

In addition to complete or almost complete specimens the Museum has sizable components of a number of aircraft as noted below.

Avro 683 Lancaster X

A nose and front centre section assembled into one unit have been put on display to give visitors a view of the cockpit area of this World War II bomber when the type came up for retirement in 1965.

Avro Canada C-102 Jetliner

The Museum holds the nose section of this aircraft, Canada's first jet aircraft to fly, and the world's second and North America's first jet transport. It was donated to the Museum by Avro Canada in 1956.

Avro Canada CF-105 Mk.2 Arrow

The Museum holds the nose section of this aircraft, which was on display in the National Museum of Science and Technology. When the aircraft was scrapped, this component was saved to serve as a pressure chamber at the Institute of Aviation Medicine, Toronto. When it had served its purpose, it was donated to the Museum in 1965. The Museum also holds the undercarriage and other smaller components of this aircraft.

Canadian Vickers Vedette

The Museum holds the nose section of the hull of Vedette G-CYVP,

Avro Canada CF-105 Mk. 2 Arrow 2 nose section and undercarriage, the largest surviving parts of any CF-105.
(NMC 71-723)

216

which was displayed from 1960 to 1967, and a considerable portion of the hull of G-CYWQ, which was salvaged from a swamp near Trenton in 1977. The Western Canada Aviation Museum at Winnipeg also holds some salvaged Vedette parts and has prepared Vedette drawings with the reference use of the NAM hull parts. It has now started the construction of a Vedette flying boat. A former Canadian Vickers engineer is supervising the project.

Northrop Delta

Some damaged parts from a crashed machine are held but they are not suitable for display or restoration.

Sopwith 7F.1 Snipe

The Museum holds the fuselage of Major W.G. Barker's VC-winning Snipe. This is the only sizable actual relic connected with any of Canada's fighting pilots, who established such a fine reputation in the first war in the air. It has been on display in the Canadian War Museum since 1967.

The remains of the Canadian Vickers Vedette hull as they were found near Trenton, Ontario, in the summer of 1977.
(NAM 12057/A.J. SHORTT)

The Powerplants

The powerplant is the heart of the aircraft and is what makes all powered flight possible with, of course, the notable exception of the recent successful man-powered flights. With dramatic improvement in every aspect of flight since the early days, it is obvious that the development of powerplants has been as impressive as that of airframes, and one would expect aviation museums, historians and writers to devote as much space to powerplants as to airframes. This is not the case though, possibly because the airframe is aesthetically more appealing than the powerplant, or the changes that have been made in airframes over the years are more noticeable. The changes in power-plants and their significance are not so obvious and thus it is more difficult to tell the story of powerplant development to the average visitor.

Some developments that have led to notable improvements in engine performance lie outside the powerplant itself: a major contribution has been the steady development of improved materials, especially new alloys of aluminum and steel. These have greatly increased in strength and resistance to high temperatures and fatigue. The difficulty facing the curator mounting an exhibit on engines is that a part made from a modern high-strength alloy of aluminum or steel looks the same as a part made from an older, less efficient alloy, yet it offers a great improvement in performance and may represent a significant technological advance.

A good example is the development of the sodium-cooled exhaust valve, which occurred in 1922. The head of the valve worked surrounded by hot exhaust gases, which made it susceptible to failure because of the high temperature. The solution was to make the valve stem hollow and fill it with sodium. The sodium became liquid at high temperature and the motion of the valve caused the sodium to move up and down, transferring heat away from the head of the valve to the valve stem, where it was conducted away through the valve guide. A hollow, sodium-filled valve appears identical to a solid valve and normally the valve is concealed within the engine. So, while a valve can be sectioned and displayed with descriptive material separately, the introduction of such a valve cannot be detected in an engine.

From the earliest days, the cooling of aircraft powerplants created two problems. One was to provide adequate cooling so that the engine did not overheat and seize up, and the other was to provide the cooling in a way that produced the minimum drag so as not to impede the flight of the aircraft. In the beginning aircraft speeds were so low that drag was not a major problem but this did not last for long.

The air cooling of early powerplants was marginal, so most early engines, and all high-powered ones, were liquid cooled. As speeds rose during World War I the drag imposed by the radiator of liquid-cooled engines began to be a serious problem. The Germans did much research on this and were the first to introduce radiators buried in the wing. In the years following the war research was carried out on both the design of radiators and their optimum location on an aircraft. This work continued right up to the introduction of the gas turbine powerplant. The optimum location for the radiator tended to vary with the type of aircraft. The introduction of a new type of ducted radiator in Britain in the mid-1930s was a major improvement in radiator systems. The entering cooling air was led through a duct that widened in cross section, reducing the air's velocity and increasing its pressure. The air passed through the radiator core, where it was heated and thus expanded. It then passed through a narrowing duct, which increased its velocity so that, when the air was expelled, it created thrust which minimized the drag and, under ideal conditions, could produce a small net gain in thrust.

Probably the most significant advance in liquid cooling was the introduction of the ethylene glycol/water mixture as the coolant in the 1930s. The increase in coolant temperature permitted a significant reduction in radiator size and consequently in weight of both radiator and coolant. Inevitably this change also brought new problems, one of which was the boiling of the coolant at high altitudes. This was solved by pressurizing the cooling system.

The static air-cooled engine, even in its modest early size, was in trouble from the beginning because the early manufacturing techniques did not permit making cooling fins of sufficient area to cool each cylinder adequately. Louis Blériot only successfully completed his historic cross-Channel flight of 1909 because a helpful rain shower cooled his supposedly air-cooled Anzani engine and permitted him to reach England. The well-documented A.E.A. experiments of 1908 record constant trouble with air-cooled Curtiss engines which only disappeared with adoption of the water-cooled Curtiss engine in the *Silver Dart*.

In fact, while a number of types of static air-cooled engines were built in the pre-World War I period, only two French makers appeared to have any success with them. One was Anzani, which marketed low-powered radial engines which continued in use on French training aircraft during

The 25 hp Anzani engine fitted to the McDowall Monoplane in the Museum's collection is the same as the powerplant of Louis Blériot's famous prototype Model XI which completed the first cross-Channel flight in 1909.
(KMM)

the war, and the other was Renault, which began making air-cooled Vee engines before the war. These were widely used by the British and French during World War I and also formed the basis of the Royal Aircraft Factory engines used during the war. Satisfactory cooling was only achieved by means of an engine-driven blower in the case of Renault and by a large ram air scoop in the case of the Royal Aircraft Factory.

To avoid the troubles experienced with the early air-cooled static engines, Louis and Laurent Seguin introduced the 50 hp Gnome Omega rotary engine in France in late 1908. This was not the first rotary engine but was the first really successful one and was widely used, setting many aviation records. More powerful rotary engines were developed by several makers and served well throughout World War I. Their development stopped at 250/300 hp because of excessive power loss incurred by the rotation of the entire engine and the increasing gyroscopic forces generated. Consequently their design and manufacture ceased about the end of the war after only a brief but noteworthy life.

Another type of aviation engine that dropped out of the picture in the early period was the two-cycle engine. A number of these engines were used before World War I, especially in North America. Some did well: the American Roberts engine was noted as being equal to the best of the four-cycle engines and was used as the powerplant for the first scheduled air service in 1914. However, the fuel consumption of the two-cycle engine exceeded that of the four-cycle engine and this became of increasing importance with the longer range and endurance being achieved by the beginning of World War I.

At the end of the war the static air-cooled radial engine was under development in both Europe and North America to replace the rotary. With advances in manufacturing processes, successful engines were built which in the mid-1920s made headlines with pioneering and record-breaking flights. Their drawback was that the cylinders protruded into the slipstream uncowled but, apparently, successfully cooled. To reduce the resulting high drag, two cowls were developed in the late 1920s. One, a narrow ring cowl invented in England, was known as the Townend ring after its developer, and the other, developed in the United States, was the wide-chord NACA (National Advisory Committee for Aeronautics) cowling. The latter was found to be the most effective, became universally adopted and gave the air-cooled radial engine a considerable step forward.

With the adoption of the NACA cowling, it was soon found that, while drag was reduced and speed increased, the engine overheated except under ideal conditions. Baffles between the cylinders were developed in England which diverted the air around the cylinders. This cured the overheating trouble with the single-row radial and, with development, baffles made possible the two-row and, finally, the four-row radial air-cooled engine.

220

The classic aviation debate through the first half century of powered flight revolved around the relative merits of the air- and liquid-cooled engine. The answer, of course, was that each had its good points and the choice had to be made after consideration of each individual application. The well-installed liquid-cooled engine usually had less drag and was apt to be chosen when speed was the major factor, but it would probably be heavier and more prone to maintenance problems because of its cooling system, as well as more likely to suffer battle damage. The air-cooled engine, with only slightly higher drag, was lighter and should require less maintenance, and consequently it has been the almost universal choice for civil use since about 1930, with the Canadair DC-4M North Star being the only exception that comes to mind.

Probably the single development that has contributed the most to engine development has been the improvement in fuel. The respected engineer S.D. Heron has written that the increase in power from the 400 hp of the 1,650 cu. in. (10 645 cm^3) Liberty 12 in 1918 to the 1,620 hp for takeoff (and 2,200 hp with water injection) of the 1,650 cu. in. (10 645 cm^3) Rolls-Royce Merlin 224 in World War II, was about half owed to fuel improvements.

It had long been known that engines ran better on fuels from one source than another but the reasons for this were not understood. Later it was found the World War I aviation fuels could be as low as 50 octane number – far below modern automobile fuels. Germany, because of its oil shortage, had to use fuels later in the war distilled from coal tar – benzene or benzol – which were actually high-octane fuels and permitted the use of high-compression ratios in some of its engines. This, in turn, permitted improved high-altitude performance and enabled some German machines to fly out of reach of Allied fighters. Studies soon after the war resulted in the discovery of tetraethyl lead's effect in avoiding engine knocking and increasing power. The US Navy was most affected by knocking and was the first to use tetraethyl lead, adding it separately in prescribed quantities upon each refuelling. In 1928 the USAAF started to use leaded (or ethyl) gasoline and shortly issued specifications for fuel by octane number.

Another development that led to better engine efficiency was the introduction of the controllable-pitch propeller during the 1930s. The conversion of the energy developed by the reciprocating engine into propulsive effect is performed by the propeller, which in the early days was of fixed pitch. The optimum pitch of the blades varies with the speed of the aircraft and the engine, as well as the air density. Inevitably the blades had to be set at an average position, optimum for some intermediate flight condition and inefficient for all others. On takeoff, where maximum power is required, the blades would be set at too high a pitch, which would hold the engine speed down and limit its output as the internal combustion engine only delivers its maximum power at maximum speed. The controllable-pitch propeller, and especially the constant-speed propeller, enabled the engine to operate at its optimum condition throughout the whole range of flight operations.

The introduction of the supercharger had a pronounced effect on the development of aviation engines. The supercharger forces a compressed fuel/air mixture into the cylinders, which makes possible the development of full power at high altitude. This enables an aircraft to cruise economically and take advantage of the lower airframe drag encountered in the rarefied atmosphere. The advantages of supercharging were realized before World War I and experimental work was carried out during the war. There are two types of superchargers, one gear-driven and incorporated as a part of the engine, and the other, known as the turbo-supercharger, driven by the hot exhaust gases and installed in the exhaust system. The first gear-driven supercharger on a production engine was installed in the Pratt & Whitney Wasp in 1927. On the first Wasp it acted more as a means to improve the fuel/air mixture distribution system than as a true supercharger, but this was soon changed on later Wasp models. A supercharger was then introduced on all radial engines to improve the mixture distribution.

Turbo-supercharger development was delayed by the lack of high-temperature-resistant steel alloys. The first production turbo-supercharger was installed with a Curtiss Conqueror engine in the Detroit Y1P-25 (later the Consolidated P-30) two-seat pursuit aircraft, the prototype of which was delivered late in 1932. The turbo-supercharger never came into widespread use but it was used by the USAAF during World War II in the Boeing B-17 Flying Fortress, Consolidated B-24 Liberator and Republic P-47 Thunderbolt, and the Germans used it in a few high-altitude Junkers Ju-88C-6C machines.

Improved machine tools over the years resulted in more intricate parts being made with finer tolerances and better surface finishes. Improved casting and forging techniques made possible better design and quality and savings in weight. One part in particular that was dramatically improved quite early was the piston. In the earliest engines pistons were of cast iron, but by about 1914 cast aluminum pistons were possible, and forged aluminum pistons were introduced about 1920. The weight reduction and improved strength of this part alone resulted in an appreciable improvement in engine performance.

There have been so many fine innovations in engine design over the years and the significant ones have been quickly adopted as standard practice. It is not possible to present even a cursory review of all these improvements here. What follows is a brief account of a few of the most significant aircraft engines.

In the beginning, the Balzer-Manly engine of 1900-1903 must be noted as a most remarkable engine. It was originally designed and built as a five-cylinder, air-cooled, rotary engine by S.M. Balzer but was not run in that form. It was converted by C.M. Manly to a five-cylinder, water-cooled, radial engine with larger cylinders, which developed 52 hp and weighed only 120 lb (55 kg), or 2.3 lb (1.04 kg)/hp. Unfortunately this impressive engine was designed as the powerplant of the unsuccessful Langley Aerodrome of 1903. Glenn Curtiss modified the Langley machine in 1914 and did succeed in flying it with its original engine.

The next engine, of course, was the powerplant of the successful Wright 1903 Flyer. The Wrights could not buy a suitable powerplant so they designed and built their own. While their design was not remarkable, it did the job it was intended to do and became the first engine to power a successful flying machine. It developed 16 hp at 1,200 rpm and weighed about 200 lb (90 lb) or 12.5 lb (5.76 kg)/hp, approximately 5.4 times the weight-to-power ratio of the Balzer/Manly engine.

In the pioneer period of aviation there was a surprising number of different aviation engines made. Probably the best of the very early European engines were the French Antoinettes of eight and 16 cylinders developing from 32 to 134 hp. They were unusual in that they had fuel injection and a cooling system in which water boiled in the cylinder jackets and the steam condensed in a condenser and returned to the engine. In the United States the Curtiss engines and, shortly, the Hall-Scott engines gave good service, and the two-cycle Roberts of four and six cylinders were said to be their equal.

The leading engine of the pioneer period was the 50 hp Gnome Omega rotary, which first appeared in France in late 1908 and came into general use in 1910. Its light weight of 172 lb (78 kg) and reliability led to its installation in many aircraft in many countries and new records in all categories were established with it.

During World War I the various rotaries of Clerget, Gnome and Le Rhône served the Allies well, with the 110 hp Le Rhône 9J probably the best all-round performer. The most significant of the Allied liquid-cooled engines was the eight-cylinder Vee Hispano-Suiza which originally appeared in 1915 as a 140 hp engine but was later developed into 150, 180, 200 and 220 hp models. It featured cast aluminum cylinder blocks and a valve gear that was enclosed and lubricated by engine crank case oil for the first time in a production aviation engine. Before this the cylinders were made separately and fastened individually to the crankcase. Cast solid blocks stiffened the whole crankcase, a feature that became standard on subsequent liquid cooled engines. While the Hispano-Suiza engines of 140 and 150 hp and the high-compression 180 hp gave good service, the higher-powered models ran into reliability problems in service and slowed aircraft production programs, the R.A.F. S.E. 5A and Spad 13 being particularly affected.

The highest-powered engine to go into production during World

This specimen 110 hp Le Rhône is installed in the Museum's Avro 504K G-CYCK. The 110 Le Rhône is considered to be the best and most reliable of the WW I rotary engines. The rotary engine's propeller is attached to the crankcase and the crankshaft to the aircraft so the whole engine revolves, unlike the radial engine which remains stationary.
(KMM)

The 360/400 Liberty was the most powerful engine to go into production in WW I. There were two versions, the 360 hp low compression naval version and the 400 hp military version. It remained in aviation use during the 1920s and was used in British tanks in WW II. It was restored for fitting in the Museum's Curtiss HS-2L.
(KMM)

War I was the 400 hp Liberty 12-cylinder engine designed by J.G. Vincent of the Packard Motor Car Co. and B.C. Scott of the Hall-Scott Motor Car Co. It was manufactured by six motor companies with Packard and Lincoln being the leading suppliers. The Liberty appeared too late to have any influence in the war but it proved to be a good engine and served in military and civil use for many years in the postwar period.

The Germans during the war relied mainly on the Mercedes family of engines produced by the Daimler-Motoren-Gesellschaft. These were liquid-cooled engines of four and six vertical cylinders and developed 100 to 260 hp with the 160 and 180 hp models being the most widely used. They proved to be good reliable engines and served the German air service well.

Following the war, and with the demise of the rotary engine, emphasis was placed on the development of the air-cooled radial engine. Successful radial engines were developed in England by Armstrong-Siddeley and Bristol and in the United States initially by the Lawrence Aero Engine Co. and later by the Wright Aeronautical Co. The Wright J-4 and J-5 Whirlwinds were the most prominent, serving widely in military and civil flying and were used in many pioneering and record-breaking flights in the 1920s, including Byrd's flight over the North Pole, Lindbergh's New York-Paris flight and Kingsford-Smith's trans-Pacific flight.

In 1926 the first Pratt & Whitney Wasp appeared and it was to prove one of the world's greatest aviation engines. It was an air-cooled radial engine and the first to enter production with a supercharger and a forged crankcase. During its lifetime the Wasp has powered first-line fighters and airline transports, won major racing trophies and made many record-breaking flights. It has powered aircraft types too numerous to mention, but among them were aircraft that did the lion's share of the work in opening up northern Canada in the late 1920s and the 1930s. It was the engine that the Pratt & Whitney Aircraft Co. was formed to make in 1925 and the last reciprocating type made by the company in the U.S. when it swung over to turbojet production in April 1951; production of the Wasp continued at Pratt & Whitney Canada until 1960. The original Wasp was developed over the years to give increased power and it led to many derivatives such as the Wasp Jr., Twin Wasp Jr., Twin Wasp, Double Wasp and Wasp Major.

In 1928 another fine but smaller air-cooled engine appeared, the D.H. Gipsy, which was designed in England by de Havilland to replace the A.D.C. Cirrus. It shortly became known internationally as a good light aircraft powerplant. Like the Wasp in the United States, it was developed through several versions, and it continued in production into the 1960s.

After World War I the first outstanding liquid-cooled engine to appear was the Curtiss D-12, which first flew in 1922. The D-12 was developed from the Curtiss K-12 of 1918, which had been designed by Charles Kirkham in an attempt to create a powerful engine with a minimum frontal area, and it incorporated features introduced in the Hispano-Suiza. The K-12 was improved upon by F.R. Porter as the Curtiss CD-12 in 1920. This in turn was improved upon by Arthur Nutt in 1922 as the D-12, in which almost all parts had been redesigned. The resulting engine had a profound effect on the liquid-cooled engines to follow. Early D-12s were primarily racing engines and easily captured the Pulitzer and Schneider trophies. They were then installed in American pursuit planes and light bombers and a D-12 development became the Curtiss Conqueror.

The influence of the D-12 was most pronounced in Britain, where it was introduced by C. R. Fairey as the powerplant of the Fairey Fox day bomber. The Fox promptly left all extant fighters behind and the Air Ministry issued a requirement for a similar British engine. Rolls-Royce took up the challenge and designed the Kestrel along similar lines to the D-12. The Kestrel was subsequently developed into the Buzzard, the racing "R" engine, and then the magic Merlin which contributed so much to the Allied cause in World War II. The first production Merlin, the Merlin I, was delivered in July 1936 and developed 890 hp for take-off. Over the years the Merlin's power was increased to 1,620 hp for take-off. It is interesting to reflect on the Merlin's background and note that its lineage, like that of a number of other engines, goes back a surprising number of years and that good engineers of at least three countries made contributions to the design that finally became the Merlin.

The liquid-cooled powerplant probably reached its ultimate power with the development of the Napier Sabre engine during World War II. The Sabre was an H engine with sleeve valves and was of most intricate design, likely the most complicated engine made. It developed up to 3,000 hp maximum take-off power.

The air-cooled radial engine continued to be developed for commercial use for several years after World War II. It reached the peak of its development with the Wright R-3350 Turbo-Compound engine, developing 3,400 hp for take-off, and the Pratt & Whitney R-4360 Wasp Major four-row engine developing 3,500 hp for take-off. It has been correctly said that "the air-cooled [radial] engine has made air transport what it is today [1959] and that it has really changed the face of the earth."

This all-too-brief story of some of the leading reciprocating engines has neglected to mention the appearance in the 1930s of a number of good horizontally opposed air-cooled engines that provided an excel-

The 125 hp Menasco D. 4 Super Pirate is the powerplant of the Museum's D.H. 82C2 Menasco Moth and is typical of the well-liked, reliable, small inline engines with both upright and inverted cylinders which gave such good service from the 1920s onwards.
(KMM)

lent source of reliable power for many light aircraft and continue to do so. Possibly the outstanding maker of these engines has been Continental Motors/Teletyne CAE, which has made four- and six-cylinder engines of this type ranging from 40 to 375 hp.

Towards the end of World War II a revolution in aircraft powerplants began with the introduction of the turbojet engine into military operations by both sides. The beginning of the revolution occurred in Germany on August 27, 1939, with the first flight of the centrifugal Heinkel He S-3b engine in the Heinkel 178 airframe. The engine was designed by Hans von Ohain. It was followed in England on May 15, 1941, by the Gloster E.28/39 with the Power Jets centrifugal W.1. engine designed by Frank Whittle (later Sir Frank). The first use of jet aircraft in action occurred on July 25, 1944, when a Messerschmitt Me 262 attacked a British D.H. 98 Mosquito over Germany. Two days later the Gloster Meteor flew its first sorties against the V-1 flying bombs over England.

Both original turbojets had a centrifugal-flow compressor, as its technology was better known than that of the axial-flow compressor, but its use resulted in a larger-diameter engine with higher fuel consumption. The Germans quickly abandoned the centrifugal compressor design and no such German engine ever entered production, but the British continued to use it in the early postwar years. Both the original engines were of modest power output, the He S3b developing 1,100-lb (500-kg) thrust and the W-1 developing 850-lb (385-kg) thrust. From such a beginning modern turbojets have developed that produce 50,000 lb (22 700-kg) thrust and more.

Since its beginning the turbojet has developed into two new forms, the turboprop and the turbofan or bypass engine. The turbojet achieves its maximum efficiency at high speed and high altitudes; to achieve good efficiency at medium speeds of 300 to 400 mph (480 to 645 km/h) the turboprop was introduced, in which the turbine drives a propeller which absorbs most of the turbine's energy with only a small portion being expended in its exhaust gases. The first turboprop to fly was a modified Rolls-Royce Derwent mounted in the nose of an Avro 683 Lancaster in September 1945, but the first engine designed as a turboprop to fly was a Rolls-Royce Dart, also mounted in the nose of a Lancaster, in October 1947. The first aircraft to fly solely on turboprop power was a Boulton Paul Balliol T.1 powered with an Armstrong-Siddeley Mamba which flew on March 24, 1948.

The two-spool development of the turbojet has two independent compressor and turbine systems, a low-pressure system and a high-pressure system, with the L.P. system supplying air to the H.P. system and thus in effect acting like a supercharger to the H.P. system and increasing the power. The first two-spool turbojet was the Pratt & Whitney J57, which first ran in 1949 and went into production in 1951.

This 16 hp Duryea engine was imported into Canada in 1906 by W.R. Turnbull, Rothesay, N.B., for use in his propeller experiments, and is the oldest engine in the Museum's collection. Although it was sold for aeronautical use, it is not known to have successfully powered an aircraft.
(NMC 80-4017)

The turbofan or bypass engine is a two-spool engine in which some of the L.P. compressed air bypasses the combustion process. In most engines the L.P. compressed air joins the combustion air aft of the H.P. turbine; the compressor is then acting as a ducted fan. The result of this gives a greater mass of air moving at a lower speed, giving greater efficiency and fuel economy. The first turbofan engine is believed to have been the Rolls-Royce Conway, which was first flown in an Avro 706 Ashton in 1955. A few turbofan engines, like the Pratt & Whitney JT3D and JT9D, discharge the L.P. compressed air directly to the atmosphere.

The next development of the gas turbine is an improved turboprop driving a multi-bladed propeller with swept-back blades made of light high-strength synthetic material. This will enable aircraft to operate more efficiently and economically at high speed and the first full scale applications of this are now being tested.

The National Aviation Museum is fortunate in having a particularly fine collection of aviation engines for such a young museum, and a list of the types held is in Appendix 4. The collection includes a number of engines of specific historic interest to Canada. While it is not proposed to review the engine collection in detail, a brief mention will be made of some engines of special interest.

The Duryea engine, made by the Duryea Power Co., of Reading, Pennsylvania, is the oldest engine in the collection. It was imported in 1906 by W.R. Turnbull for use in his propeller experiments of 1906 and 1907. While it was marketed for aeronautical purposes, the engine is not known to have powered a successful flying machine and it is known to have given considerable trouble to Turnbull in trying to carry out his experiments. Nevertheless, it is of interest both as an early engine and as a relic of Turnbull's early aeronautical work.

Next is the Curtiss *Silver Dart* engine of 1908, which powered the A.E.A. *Silver Dart* when it made the first flight by a powered heavier-than-air machine in Canada on February 23, 1909. It is a fine relic of this historic event. In addition, it is only the second water-cooled engine made by Curtiss, the first being a four-cylinder but otherwise identical engine supplied to Capt. T.S. Baldwin for his airship. These two engines are also the only Curtiss engines known to have used concentric valves and the Museum's specimen is the only survivor.

The Museum is fortunate to have the first Canadian-built aircraft engine, the six-cylinder air-cooled Gibson engine of 1910. Actually W.W. Gibson had built an even earlier four-cylinder engine but it was unsuccessful. The engine was made in Victoria, except for the crankshaft, which was made in New York, and the crankcase, which was cast in Seattle. Gibson used the engine to power his Twinplane in 1910 and his Multiplane in 1911.

This 40 hp Curtiss engine powered the A.E.A. Silver Dart *on its first, powered, heavier-than-air flight made in Canada on February 23, 1909. It was the second Curtiss water-cooled engine made and one of only two Curtiss engines with concentric valves.*

(NMC 80-4016)

The 50 hp Gibson was the first successful aircraft engine to be made in Canada. This one was used to power his two aircraft, the Twinplane in 1910 and the Multi-plane in 1911.

(NMC 80-1671)

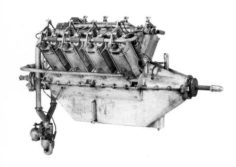

The successful 1911 80 hp Curtiss O engine was further developed in 1912 with an improved valve gear as the Model OX. The OX was further developed during WW I to the well-known OX-5. This specimen's history is undocumented but it seems likely that it powered the first Canadian flying boat made by Percival Reid at Montreal in 1914.

(NMC 80-1696)

Another engine that has an interesting history is the Curtiss O engine which was acquired by the Canadian War Museum in Quebec, north of Montreal, in the belief that it was a Curtiss OX-5 engine of World War I. The Curtiss O was a fine pioneer engine introduced in 1911 and the ancestor of the well-known Curtiss OX family of engines which began in 1912. Two Curtiss O engines are known to have come to Canada in the pioneer days, one brought to Vancouver in 1912 by W.M. Stark in his Curtiss pusher and later used by T.F. Hamilton in the second Hamilton biplane of 1916, and the other imported by P.H. Reid of Montreal in 1914 to power his Reid-Morgan flying boat. This aircraft flew in late October 1914 and was the first Canadian-built flying boat. By geographical association it seems probable that the Museum's engine was its powerplant, but it is unlikely this can be proven.

All the many engines built in the pioneer period are interesting. Besides those already mentioned, the collection includes the following engines from this period: Anzani (fan type) 25 hp, Anzani radial 40/45 hp, Curtiss L, Dutheil and Chalmers 20 hp, Elbridge Aero Special four-cylinder, Gnome Omega, Kirkham B-6, and Roberts 6X. Some are now restored, and it is hoped that the rest can be restored for display in the near future.

The Museum is also fortunate in its collection of World War I engines, especially those of British and French origin, but also some of German design. Some of the rotary engines have been used to fly a few of the Museum aircraft. The collection also includes significant engines of the between-the-wars period as well as many of the World War II powerplants.

The Museum has several engines of special Canadian interest from the post-World War II period. The Avro Canada Chinook engine, the first Canadian turbojet, was a donation of Avro Canada. It was built to prove the theories of its engineers, which it did, but it was never flown, contrary to some statements. The Museum holds examples of the first Canadian turbojet to go into production, the fine Avro Canada Orenda, together with an Orenda Iroquois, which was designed to power the Avro Canada Arrow 2 but unfortunately did not go into production. There is a cut-away example of Canadian Pratt & Whitney's fine PT-6, the first Canadian turboprop and an engine that is widely used internationally in general aviation aircraft.

Most major museums have a restoration department to restore, repair, preserve and conserve the artifacts that fall under their jurisdiction. This work, while of great interest, is seldom seen by the general public, although articles appear from time to time dealing with the restoration of a specific item.

Each museum develops the knowledge and skills necessary to carry out restoration on objects in its field. So one museum becomes expert in restoring paintings, another, statues or urns, and still another, wooden ships that have been buried or sunk for hundreds of years. The science of aeronautics is a young one, with the first unmanned balloon ascent taking place in 1782. Few relics of the balloon era have survived, so aeronautical artifacts usually date from the beginning of the heavier-than-air era at the start of this century. As they are relatively young in comparison with the relics of early civilization, it might be assumed that there would be few problems with aeronautical artifacts. However, this is not so, for while they are remarkably strong to resist the forces of flight, they are also light and fragile and easily damaged in the hands of unknowledgeable persons. Consequently their survival rate is not high, and surviving items usually require considerable work to bring them to condition suitable for display.

Restoring aircraft components presents difficulties not usually encountered elsewhere. First, probably no other object of human origin has embodied so many different types of material. Most types of wood have been used in airframe construction, usually in their highest grades, which are often difficult to obtain today in appropriate sizes and lengths. Most metals have been used, usually in the most advanced alloys available at the time of their construction. A wide variety of fabrics including linen, cotton, silk and synthetic fabrics, and varnishes, paints, enamels and other finishes have been employed, along with an assortment of cements and adhesives.

To restore aircraft correctly, the characteristics of all the different materials must be understood, together with the proper methods of working with and storing them. The restoration department can usually make or repair most parts that are required but it is not practical to equip the workshop with specialized tools that might be needed only occasionally. Sometimes a part will be encountered that cannot be made with the general-purpose tools in the museum's shops, and, in these cases, it is necessary to call upon outside help.

In most cases old airframe parts can be restored, repaired or, if essential, replaced by skilled and diligent work; the difficult job is finding original equipment that has been removed or replaced by more modern equipment over the years. In these cases, all that can be done is to make extensive inquiries to try to locate the necessary items. Most museums welcome the donation of obsolete aviation equipment of all types, which almost always will in time be put to good use. Tires, for example, have become a very difficult item to replace. Original tire molds are seldom available and the cost of making a new single set of tires is usually prohibitive. As a result, many tires on older museum aircraft, including those of the National Aviation Museum, can no longer bear the weight of their specimens and are therefore suitable for display only.

The basic practices followed by the National Aviation Museum in its restoration programs were established in their first restoration, the Curtiss JN-4 (Can.), carried out from 1962 to 1964. These may be summarized as follows: parts should be preserved rather than restored, restored rather than replaced, and replaced only when essential. Thus, original parts, including wooden parts, are used wherever possible. In some cases a piece of new material will be spliced into an old part to preserve as much as possible of the original. The original type of material is used in the replacement part when necessary, and all details of the original construction are followed even if more modern short cuts are available. Replaced parts are marked for future reference. Modern paints and varnishes are used, as are modern adhesives which are a great improvement over the old glues. Fabrics to cover the wings and tail surfaces are linen or cotton as on the originals, not modern synthetic fabrics.

The only exception to the use of original material other than the paints and varnishes mentioned above is the substitution of the modern methyl methacrylate for celluloid or pyralin as windshield and window material in early aircraft. This is because the early material aged and became yellow and opaque while the modern plastic remains clear and is not otherwise distinguishable from the original product. Also, the A.E.A. *Silver Dart* used rubberized fabric as its covering material, while the Museum's reproductions, built by the RCAF, have doped fabric.

Most restorations require considerable research to ensure a

satisfactory and historically correct result, for any or all of the following: damage to the specimen; missing parts or components; modifications during the specimen's service life; unknown original colours and markings; and construction processes and skills that are no longer in use. When such conditions exist, locating the necessary information can be a difficult and time-consuming job. Usually the original drawings no longer exist, so they cannot be used for reference. The Museum's fine reference library and photo collection are invaluable for this work.

The accounts that follow deal with restorations done at the Museum, not those done by others for the Museum. For a description of a fine restoration by Jack Canary, see Appendix 2. The projects have varied greatly in size and complexity from the little Aeronca C-2 and Sopwith Pup to the large and complex Curtiss HS-2L. The latter is the largest restoration project carried out by the Museum so far and it will probably be so for some years to come.

Before the Museum started on its first restoration project, standards were set to ensure a high quality finished product. First, the colour values of all finishes had to be recorded so they could be duplicated; similarly the style, size and location of all markings had to be recorded. Notes were made of all markings, dates, etc., found on the various parts, as they are helpful in establishing the aircraft's identity and early history. All parts removed are to be retained for future reference. All fabric must be removed in large pieces and retained for reference, as they are invaluable in determining the original covering practices, stitching type and pitching, etc. On completion of the restoration the parts and fabric are usually retained for future reference. A record of all complicated assemblies had to be made before disassembly so that reassembly to their original state may be done with confidence. This is usually done photographically.

Finally, a photographic record of all components should be made where feasible, both as a record of what was done and as a record of the specimen's internal structure for historical purposes. This latter procedure is historically important, as the specimens restored are usually rare and sometimes unique.

Curtiss JN-4 (Can.) "Canuck"

The specimen as obtained by the Museum in 1962 was basically complete. The wings still had the original 1918 fabric on them, but all other components had been stripped. It had been in civil use and the instruments were not original except for the oil pressure gauge; other items were missing or damaged.

When the specimen was obtained, the Museum had no workshop

The complete flight control system is shown here reassembled on the cockpit floor boards and ready to be placed in the aircraft.
(KMM)

RIGHT
Bill Merrikin reassembling a portion of the front fuselage of the JN-4 Canuck.
(KMM)

capable of housing it. Nevertheless, because of its historical importance to Canada, it was felt that a way must be found to start its restoration. A narrow area at the back of the Museum at Uplands was made available, and while most components were carried in through the twisting hallways, the fuselage had to be hoisted in through a second-storey window.

After being displayed at Rockcliffe on Air Force Day in June 1962, the specimen was trucked to the Museum and restoration started. We decided to restore the machine with the fabric removed from one side as a result of favourable comments when it was displayed with its structure visible. At that time no drawings were available to the Museum, but a stress report contained some useful information and a large number of photographs proved invaluable. A large-scale Canadian Aeroplanes-made model showing all structure in detail also provided useful information.

The fuselage was completely disassembled and restored in two stages. First, the portion from the rear cockpit aft was done, and then the forward section. All metal parts were cleaned and repaired or plated as required. All wood parts were cleaned and varnished. Only three wood parts required replacing – the right horizontal seat bearer and the large left upright member at Station 3, along with the transverse support for the engine bearers at that station. During this work the American serial 39158 was found on the upper right longeron in the

232

LEFT
A new turtle deck for the rear fuselage is being assembled by Arthur Walker for the Canuck.
(KMM)

RIGHT
This photo provides a good view of the wing bracing and aileron control cables. The aileron control system was unique to the Canuck in the JN-4 family and is an excellent and reliable recognition feature.
(KMM)

rear cockpit, establishing the original identity of the machine for the first time.

The cockpit flooring was cleaned and revarnished, all flight controls were removed and re-plated, and a new front control stick socket and stick were made to replace the crude ones made by some previous owner. New engine cowling panels were made to replace the badly worn and cracked original ones. The rear fuselage turtle deck, of later design and in very poor condition, was replaced by one of earlier design based largely on the detailed model.

A new instrument panel was made to the correct layout, and all original-type instruments, lights, etc., were located except for the airspeed indicator for which a dummy was made. New throttle controls were fitted in the front cockpit; Deloro Stellite kindly provided the casting. The seats were repaired and re-upholstered. A correct style of seat belt was donated by Ed Carlson of Spokane, Washington, which served as a pattern for new ones and all-new webbing was installed in both belts, with needed castings again made by Deloro Stellite.

The tail surfaces were generally in good condition. The wood members required only minor repairs, cleaning and revarnishing, and the tubular members were treated internally with linseed oil as a preservative. New elevator trailing edges were required due to corrosion, and as the Museum could not form them in the required lengths, Canadair Ltd. generously supplied them.

The undercarriage required only cleaning and refinishing, but the spoked wheels were rusted, and to avoid complete disassembly Canadair kindly cleaned them as a unit in their ultra-sonic cleaner.

The wings, in general, required only minor repairs. However, the aircraft had been in an accident at some time, which resulted in the root fitting of the left upper front spar being pulled out of the spar and the end box rib being badly damaged. As only very crude repairs had been made originally, spar repairs and a new end rib were required. All-new trailing edges were required. All ailerons were warped so four new ones were made.

The OX-5 engine obtained with the machine was overhauled and placed on separate display, and a second OX-5 was overhauled and placed in the aircraft. A donated Canadian Aeroplanes-made propeller was refinished and installed to replace the American-made one obtained with the machine.

As doping was not permitted in the Museum building, the covering and doping were done under contract by Personal Plane Services to the Museum's specifications. The method of applying the fabric to the JN-4 (Can.) fuselage was unique and it took considerable research before all its details were determined. The aircraft was finished as a machine of the 85th Canadian Training Squadron bearing its Black Cat insignia. The colour of the cowling was researched and a particular shade of green was selected as specified by a former senior Canadian Aeroplanes employee. However, it has since been determined that this colour is incorrect, and some day the specimen will be refinished in khaki brown. The restoration was completed in May 1964.

Fairchild FC-2W-2

It was hoped to start the restoration of the FC-2W-2 immediately after completing the Curtiss JN-4 (Can.), but the Uplands workshop could not accommodate the machine. Shortly, space was rented in a warehouse on Terminal Avenue in Ottawa and about half of the restoration was done there before moving to Hangar 68 at Rockcliffe the following year.

The aircraft had suffered an engine failure on a delivery flight in 1943 near Philadelphia. In the ensuing forced landing, it impacted heavily on the right undercarriage, destroying an undercarriage casting, ripping an undercarriage fitting from the fuselage, pulling some fuselage welds and causing some slight fuselage distortion. A tree penetrated the leading edge and front spar of the right wing just outboard of the wing strut attachment fitting. In addition, something hit the tailplane leading edge, causing a clean break through it. Restoration involved a considerable repair job, but this time, fortunately, most drawings were available.

In order to straighten and reweld the fuselage, all the secondary wood structure which was mostly in the cabin area, had to be removed. The necessary breaking of the glued joints required replacing almost all of this structure with new material except for the structure on the cabin roof, which was removed as a unit. The tubular structure was

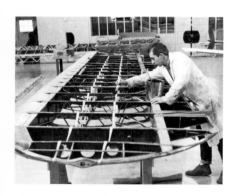

A Museum employee is putting finishing touches on the wing of the Fairchild FC-2W-2 prior to covering with fabric. The white fabric tapes strung between the ribs stabilize them and prevent twisting and warping.

(PAC PA-134309)

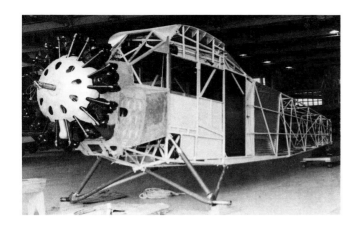

straightened, a new undercarriage fitting made and the fuselage tubes were filled with hot linseed oil and then drained as a preservative measure. In rebuilding the secondary wood structure, it was necessary to recreate the original baggage compartment aft of the cabin, which at some time had been made into a camera compartment.

In the pilot's compartment a new instrument panel was required and new instruments had to be found and fitted. A new pilot's seat and elevator trim wheel had to be made, along with moulding a new rubber grip for the control stick. By good fortune it was possible to obtain cabin lining material identical with that used originally.

A wood pattern had to be made for the fitting at the junction of the undercarriage legs. It was cast by Deloro Stellite and machined by the Museum. The sheet metal fairings covering the undercarriage legs were missing so new ones were made to drawings. The tail surfaces, made of steel tubing, were in good condition and required only cleaning and refinishing, except, of course, for the break in the leading edge, which had to be spliced.

A new section of the right front spar had to be spliced in, extending from just outboard of the strut fitting to the tip. The plywood leading edges had to be replaced completely on the wings owing to glue deterioration, and as usual there were minor rib repairs. Fortunately the ailerons and flaps for the folding wings were in good condition and required little attention. One or two of the wing struts were badly pitted from lying on damp ground. As streamlined tubing of that size was not available, it was decided, with regret, to clean them and restore them to display condition only.

Since both the propeller and engine had been removed from the airframe after its forced landing, it was necessary to locate these items. The USAF Museum had a Pratt & Whitney Wasp C surplus to their needs, which they loaned to the Museum and later, generously,

LEFT
In this view the fuselage structure has been completely restored, wood fairings have been added and the cabin lined and ready to receive its fabric covering.
(KMM)

RIGHT
In applying the G-CART identification to the Fairchild, these markings were faithfully followed down to the missing hyphen between the G and C, apparently inadvertently omitted on the original.
(KMM)

Instrument panel of the Fairchild FC-2W-2. On the left are the throttle and spark advance levers; the lever on the right controls the cooling air vents in the nose cowl.
(KMM)

235

The interesting Fairchild wing folding arrangement can be clearly seen in this photo. To fold the wing, a catch releases the hinged trailing edge flap which is then folded onto the top of the wing as shown. While one man secures the wing by holding the wing struts, another pulls the lever visible at the wing leading edge. This releases the locking pin and the wing folds back of its own weight while restrained by the other man.

(KMM)

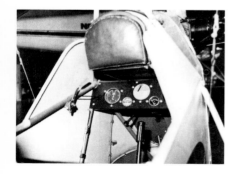

The very simple cockpit of the Aeronca C-2.

(KMM)

donated. It was restored to display condition, and this involved casting five new rocker box covers, which were made by Deloro Stellite. Obtaining a ground-adjustable propeller proved difficult but one was finally purchased from an American propeller shop in good display, but not airworthy, condition.

A hazardous condition was found in the course of restoration that would have caused trouble, quite possibly serious, if the machine had continued in service. The baffle in the centre fuel tank had cracked along its lower flange where it joined the tank bottom. Its sharp edge had gouged a slot several inches long in the tank bottom to the point where it was almost completely through the bottom. When the inevitable failure occurred, the contents of the tank would have been very quickly released into the cabin, which at best would have been troublesome and at worst catastrophic. As there was no intention of flying the restored aircraft, the tank was not disassembled and repaired, but a record of its condition was placed in the aircraft file.

It was decided to finish the aircraft as a Canadian-registered FC-2W-2 and to select the serial number nearest to the specimen. This proved to be G-CART, registered in July 1928 to Canadian Transcontinental Airways Ltd., whose construction number was only two away from that of the specimen. The restoration was completed in May 1966.

Aeronca C-2

Following the restoration of the FC-2W-2, the Museum staff restored several engines and a D.H.82C wing for an intended structural exhibit. However, when the Aeronca C-2 arrived at Rockcliffe in January 1967, its diminutive size and unusual design aroused the interest of the RCAF pilots who were preparing to fly some of the Museum's aircraft. They asked if it could be flown also.

The aircraft had to be test flown about the end of May, and a quick check of the airframe indicated that it was possible, although time was short for the work that would have to be completed. Because it involved only recovering and minor repairs, it is probably more realistic to describe the job, like the later one on the Sopwith Pup, as a recovering and reconditioning operation rather than a full-fledged restoration.

The two major questions concerned the streamlined flying and landing wires and the Bosch magneto. The wires had to be made to order in the United States and the delivery date left no margin for any delay. The RCAF kindly took care of the magneto, sending it to Germany for overhaul, and it arrived back in good time.

The minor repairs required to the airframe caused no difficulty. It was decided to finish the aircraft as the first Aeronca C-2 that came to

The original CF-AOR, the first Aeronca C-2 to come to Canada, is shown in its second finish scheme at Cartierville, Québec, in 1935.
(KMM)

Canada but in its more interesting second finish scheme; fortunately its civil registration was not in use and was obtained. The work was completed in good time but the streamlined wires were delayed, arriving just in time to permit a test flight on June 9, the day before the Air Force Day display.

The Seagull following a trial assembly at Rockcliffe in June 1968.
(NAM 8022)

Curtiss Seagull

As related elsewhere, no restoration work was carried out at Rockcliffe from the completion of the Aeronca C-2 until work started on the Curtiss Seagull in October 1970. The Seagull arrived at Rockcliffe in May 1968 and a trial assembly was made in June. The machine was found to be complete except for the bracing struts from the lower main wing roots to the hull and the stabilizing struts from the rear of the engine bearers to the upper wing.

Inspection of the hull showed that the plywood covering had deteriorated and required replacing. This was not surprising and is typical of older plywood made before modern adhesives became available. This would be a major restoration project and the Museum,

The hull joints are being cleaned and reglued. (NAM 10415)

The finished Seagull hull with its varnished mahogany plywood and multitude of bright brass screws gives the impression of a piece of fine furniture. (KMM)

The Seagull's cockpit was unusual in that it had wheel control fitted on the left seat but stick control on the right. The reason may be that its first pilot, an ex-USN officer, was used to the wheel control, while its second pilot, a USAS officer, was used to stick control. (KMM)

understandably, was reluctant to undertake it as the aircraft was then only on loan. An exchange was subsequently arranged for the Seagull and restoration began in October 1970. The job was started with much enthusiasm, as flying boats had carried the lion's share of early northern aerial operations in Canada and this was the first early example obtained by the Museum.

Removal of the plywood, which was screwed to the framework, revealed a structure consisting of an ash keel and longerons and spruce trussing, a far simpler arrangement than on the Curtiss HS-2L and other early Curtiss flying boats. All the frame members were in good condition, but the glue had vanished from the joints, possibly as a result of the aircraft's period in the tropics. The whole framework had to be disassembled, cleaned and glued, and reassembled. Also, all the screw holes in the framework had to be filled with plugs in preparation for installing new plywood. This operation was not difficult, but the large number of holes made it a tedious job.

Obtaining the proper mahogany plywood for the hull proved the most difficult task in this restoration. The hull had been made using long plywood sheets that enabled the hull sides to be enclosed with only one splice. These sheets were no longer made in Canada and in spite of numerous inquiries in the United States and Europe no supplier could be found. Regretfully, the Museum had to complete the hull using smaller sheets with two splices and these were secured to the framework with the same multitude of small screws as had been originally used.

The missing struts posed no problem, as their construction details were confirmed by checking another extant Seagull and a similar Curtiss MF flying boat. There were no problems with the Curtiss C-6 engine restoration, as no parts were missing, and it needed only the usual disassembly, cleaning, refinishing, and reassembly. However, it was found that the cylinder head had been installed in reverse, which must have led to cooling problems.

The wings and tail surfaces required a great deal of work, as they had to be completely disassembled and cleaned; many pieces had to be spliced in, and new cap strips and metal trailing edges had to be made and installed. Everything then had to be reassembled, reglued and varnished before covering.

The standard factory finish for Seagulls was clear doped wings and a clear varnished mahogany hull. However, for use at the equator the Hamilton-Rice Expedition had their machine overcoated with aluminum as protection from the sun. It bore the name *Eleanor III* on each side of the bow, with a Brazilian flag under one wing and the American flag under the other. The original painted flags and names have been preserved and are on display near the aircraft. The specimen was finished with aluminum doped wings and a clear varnished hull. It was completed in February 1974 and went on display officially on March 12, 1974, after a brief ceremony.

Sopwith Pup

In October 1973 the Sopwith Pup was acquired, and while it was a relatively new reproduction, it needed recovering and the inevitable tightening up of the airframe; in addition the trailing edge of one control surface had to be replaced. So, as in the case of the Aeronca C-2, it was less than a full restoration and was carried out while the Seagull was being completed.

To give it a finish scheme more colourful than the original, it was completed as aircraft B2167 of No.66 Squadron, RFC, bearing the squadron letter "L." It was first flown by the Museum in the summer of 1974 and appeared at air shows at Winnipeg and at Abbotsford, British Columbia.

Fokker D.VII

In February 1971 while the Seagull was being restored, the Museum was fortunate to obtain an original Fokker D.VII, one of only seven believed extant. We had desired to establish its original identity, as there were small differences between the machines produced by the three different makers of the type, and then establish, if possible, the type of lozenge fabric that was used to cover it originally.

The Fokker Werke number D.VII 3659 was found on the upper wing, which (assuming the wing was the original supplied with the aircraft by Fokker) established it as being Fokker-built and having borne the German Air Force serial 10347/18. This was interesting as it is believed to be the only Fokker-built D.VII extant. The serial showed it to be one of the last D.VIIs built and among those seized at the factory by the Allies in 1918. Also of interest is that serials 10349/18 and 10350/18 were brought to Canada as war trophies (see Appendix 1) and serial 10348/18 was with the US Air Service. Photographs of all three of these machines assisted restoration, although no photograph of the Museum's specimen in its original finish has been found.

In 1917 the Germans and Austrians were forced by a shortage of pigments to resort to dyed fabric to camouflage their aircraft. The Germans made two different fabric styles, a four-colour and a five-colour pattern. The patterns consisted of a series of pentagons, and the fabric has become widely, although incorrectly, known as lozenge pattern fabric. In addition, each style of fabric was made in a darker and a lighter shade, with the darker shade used on the upper surfaces and the lighter one on the under surfaces.

While no fabric was on the surfaces of the specimen when it was

Installation of the Curtiss C-6 engine in the Seagull.   (KMM)

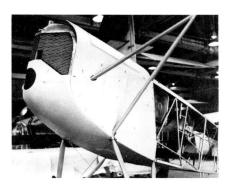

A complete new radiator had to be made for the Fokker D. VII along with new engine cowlings as seen here. (KMM)

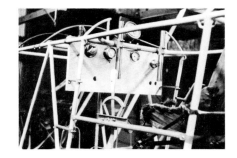

The Fokker D. VII's instrument panel fitted for a trial assembly.   (NAM)

The D. VII's upper wing after restoration; the wire trailing edge has not yet been fitted.

(KMM)

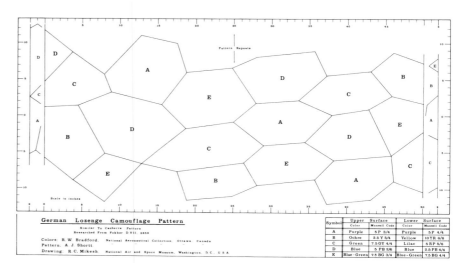

German Lozenge Camouflage Pattern

Similar To Canberra Pattern
Researched From Fokker D.VII, 3056

Colors: R. W. Bradford,   National Aeronautical Collection, Ottawa, Canada
Pattern: A. J. Shortt.
Drawing: R. C. Mikesh.   National Air and Space Museum, Washington, D.C., U.S.A

Symbol	Upper Surface		Lower Surface	
	Color	Munsell Code	Color	Munsell Code
A	Purple	5P 3/4	Purple	5P 4/4
B	Ochre	2.5 Y 5/4	Yellow	10 YR 6/6
C	Green	7.5 GY 4/4	Lilac	5 RP 5/6
D	Blue	5 PB 2/8	Blue	2.5 PB 4/4
E	Blue-Green	7.5 BG 3/4	Blue-Green	7.5 BG 4/4

RIGHT

Drawing showing the pattern and colours of dyed fabric covering, determined as related in the text.

(NASM)

The D. VII's lower wing as received. The damaged plywood leading edge is readily visible here; the rib plywood had deteriorated.

(KMM)

received, strips of the covering material used as a protective wrapping on the spars were found. The strips had been protected from light over the years and the original colours had not faded. Enough of the strips were obtained to piece together, jig-saw puzzle fashion, a sample of the original fabric. Assistant Curator Fred Shortt laid out the geometric pattern and Curator Bob Bradford carefully established the correct colour values. Then three museums, the National Air and Space Museum in Washington, the RAF Museum in Hendon and the National Aviation Museum, joined together to have a company make fabric to the established pattern for their own use. Unfortunately, the batch for the National Aviation Museum was faulty and was rejected, so the Museum does not yet have the proper fabric for its D.VII.

The Museum was most anxious to have the D.VII restored and, being busy with the Seagull, entered into a contract with the late Joe Defiore of Troy, Michigan, for its complete restoration. Defiore had been interested in the D.VII for a long time and had collected much information on the type with the intention of making a reproduction for himself. Unfortunately his work on the Museum's specimen proceeded more slowly than expected and the machine was returned to Rockcliffe in March 1974 only partly restored.

Defiore had completed an excellent radiator, cleaned and repainted the tail surfaces, replaced an upper fuselage longeron, cleaned and repainted the balance of the fuselage structure, and fitted a new seat and instrument panel. He also made a new undercarriage to replace an undercarriage of different design, which had been substituted some-time earlier in the aircraft's life. He also made and fitted a set of engine cowling panels along with their special fasteners. However, it has since been discovered that some panels differ in some details to those fitted to Fokker-built D.VIIs, so these will have to be corrected or replaced.

240

Fokker D. VII 10348/18 at Coblenz, Germany, May 9, 1919, with Lt. Everett R. Cook and Maj. Reed M. Chambers, USAS.

(USAF MUSEUM)

The Museum's D.VII had a Hall-Scott L-6 engine installed when received, but the Museum was able to obtain in an exchange a zero-time Daimler Mercedes D.IIavu engine which had come to Canada in 1919 in one of the war trophy D.VIIs (see Appendix 1). An original D.VII propeller was generously donated for the aircraft and the necessary instruments had also been obtained. This meant that all items were now on hand to complete the restoration to its original 1918 condition, except the wheels and lozenge fabric.

The Museum started work on the wings and found that while the spars were sound, the ribs and other wood parts were suitable only for use as patterns and had to be replaced. When the Museum stopped work on the wings in September 1974 to start the Curtiss HS-2L restoration, all wood parts had been made or restored and fitted.

Work remaining to be done may be summarized as follows: install internal wing bracing and wire trailing edges, install flight and engine controls, complete cockpit details, modify engine cowlings and make new spoked wheels. It is planned to exhibit the aircraft uncovered for some time before adding the fabric. It will then be finished as a factory-new machine identical to serial 10348/18 as shown in the photograph, except for its 10347/18 serial.

Messerschmitt Me 163

The Curtiss HS-2L was the next restoration project to begin, but while it was in progress the Me 163 was started and completed.

The National Museum of Science and Technology had been anxious to obtain the loan of a Convair Atlas missile for outside display and the USAF Museum wanted the loan of an Me 163. The Museum had two Me 163s, one of which had been in good condition and had

241

The fuselage of the Messerschmitt Me 163 under restoration at the NAM.
(NAM 12587)

This sketch was found on the inside of the Me 163, apparently made by an unhappy impressed French worker. The sketch is of a building, probably the factory, with the notation Manufacture ferme *(Plant closed) and underneath the message* Mon coeur est en chômage *(My heart is not in it).*
(NAM)

been refinished by the Canadian War Museum in the marking of 1 JG/400 during 1966-67 and put on display in the War Museum. The other specimen had been stored outside at St. Jean, Quebec, and was badly weathered when rescued by the War Museum in 1958. The specimen was then stored inside but no work had been done on it. It was agreed that the National Aviation Museum would restore it and place it on long-term loan to the USAF Museum. In exchange a Convair Atlas would come to the NMST.

Restoration work started in September 1976 and a quick survey showed the following items were missing: instruments and instrument panel; cockpit controls; tail wheel strut, wheel and tire; plastic cockpit canopy and bullet-proof glass; generator propeller and fairing; numerous fairings and hatch covers. The wooden wings were in poor condition from long exposure to the weather, with the left wing considerably worse. Fortunately the other Me 163 nearby made it easy to obtain the necessary information to duplicate missing parts.

The instrument panel was made, and fortunately almost all the missing instruments were found at the National Research Council, so only two dummy instruments had to be made. A mould for the plastic canopy was formed commercially by contract. The Museum formed the bullet-proof screen from plate. The rest of the missing parts were made by the Museum, but a wooden tail-wheel tire was substituted for the original rubber one.

The wooden wings had to be entirely reskinned and about two thirds of the ribs had to be replaced. The sheet metal fuselage had to be stripped of all components, and everything was cleaned and repaired and reassembled. Similarly, the undercarriage needed cleaning and refinishing as did the Walter rocket engine. All jacks, control systems, etc., were restored to working condition.

Several items found during the Me 163 restoration showed the consequences of employing impressed foreign labour, as was done in Germany during the later war period. All glue in the wings was found to be completely ineffective; Museum staff thought that soap had been mixed with it. Had the aircraft been flown, it would probably have suffered a structural failure. The large fuel tank located behind the pilot to contain hydrogen peroxide (*T-stoff*), a highly corrosive substance which, when mixed with hydrogen hydrate (*C-stoff*), ignited and powered the rocket motor, had been sabotaged. This had been done by placing a small stone between one of the tank support straps and the tank itself. The intention was that, when the tank was filled with fuel, the stone would penetrate it causing a disastrous leak.

The completed aircraft was finished in its original factory paint scheme and shipped on loan for exhibit to the USAF Museum in November 1978. It was returned to Rockcliffe for exhibit upon the opening of the Museum's new building.

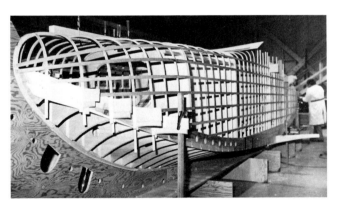

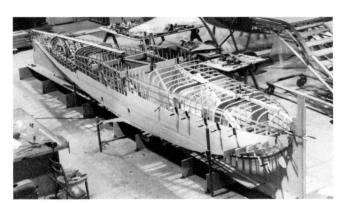

Curtiss HS-2L

To date, the largest and longest restoration project undertaken by the Museum is that of the Curtiss HS-2L. The magnitude of the job resulted partly from the large size of the machine and partly from the need to reconstruct a complete hull. While a portion of the original hull had been salvaged, as related elsewhere in this book, the condition of the wood was not suitable for restoration. In addition, the Museum wished to retain this portion in its salvaged condition as a historical exhibit.

The project began in September 1975 with construction of the hull. It is not practical to give a detailed account of this complicated operation here, but an abbreviated description with the accompanying illustrations provides a good idea of the hull's construction.

This construction was typical of the early Curtiss flying boats, and the HS-2L was the last one to use it before it went out of production in 1918. First, the contour of the keel was cut in a series of plywood sheets

TOP LEFT
The Curtiss HS-2L hull on November 25, 1975, about two months after construction started. (KMM)

TOP RIGHT
By May 20, 1976, the six mold frames or bulkheads had been installed along with the light ash frames. (KMM)

BOTTOM LEFT
By November 1976, the hull planking had also been started. (KMM)

BOTTOM RIGHT
In November 1976, the seam battens, running fore and aft to receive the hull planking, were installed. (KMM)

To double plank the hull bottom, it had to be inverted. Two plywood forms cut to circular shape were secured around the hull and the hull rolled over. (NAM 12850)

An interior view of the HS-2L cockpit. The lever at upper right is the throttle. The two control wheels are mounted on a wood yoke which bridges the legs of both pilots, a typical control arrangement for large flying boats. At the bottom of the pilot's panel is the manual fuel pump which is used until sufficient speed is attained for the air driven pump to take over. The numeral A-1876 is the USN serial of G-CAAC.

(KMM)

* These were added by the Curtiss Co. to their flying boats beginning in 1914 to increase their planing area and improve their take-off characteristics. They were called "fins" by Curtiss but such additions are usually termed sponsons today.

which were set vertically on the floor; this formed the support or "jig" during the assembly operation. Then the steamed green ash keel pieces, one forward and one aft, were clamped in place. Next the fabricated floor frames, or floors, were placed at the prescribed intervals along the length of the keel and then the chine stringers were fitted along the outer ends of the floor frames. The six assembled mould frames, or bulkheads, had to be placed in position on the hull. Next the green ash frames had to be formed to the varying hull cross sections and set in place at intervals varying from about 6 to 8 inches (15.2 to 20.3 cm) along the hull. Temporary battens were set in place from bow to stern and the frames all brought into alignment with one another to produce smooth hull lines. These battens were then removed and the seam battens fastened in position from bow to stern, and pitched from one another to suit the width of the hull planking. Then filler strips were fitted to each frame between the battens to bring them level with the seam battens.

The upper portion of the hull was planked with pine planks 5/16 inch (7.9 mm) thick. The structure of the fins* (or sponsons) was added to each side of the hull from the bow to the step, and following this the whole hull was inverted to permit planking the bottom. The bottom planking was of 5/32 inch (3.9 mm) mahogany and was installed in two layers. The first layer was at 45° to the keel and a coat of marine glue was applied and fabric laid over it. This was followed by another coat of marine glue and the second layer of planking was applied parallel with the keel. The hull was then placed upright, the upper surfaces of the fins were planked, and the cockpit coamings, already formed to shape, were fitted into the hull. The hull was then complete.

The tail surfaces were from the Pacific Marine Airways HS-2L, and while they were complete, their age and previous service made considerable work inevitable. They required disassembly, cleaning, rib repairs, one or two new ribs and new trailing edges: some of the metal parts of the rudder were corroded and required repairs. All were put in good shape in quite short order.

The wings, also from the Pacific Marine Airways machine, were another matter mainly because of their size. The upper wing spanned 74 feet (22.6 m) and was made in five pieces, while the lower wing was 64 feet (19.5 m) in span and made in six pieces. All of these wing sections required disassembly, rib repairs, and often some rib replacements; one section required a new piece spliced into its solid spruce spar. All trailing edges required replacement and all four ailerons required repairs. All struts required only cleaning and refinishing, but all-new bracing cables had to be made up and fitted. Needless to say, the restoration of the wings, struts and cables was a time-consuming job.

In addition to the airframe components, the 12-cylinder Liberty engine had to be disassembled, refinished and reassembled, and its plumbing, oil tank and radiator shutters made and installed. The new oil tank and radiator shutters were based on some skimpy and much-corroded remains and, of course, photographs. The radiator itself, salvaged from the Ontario Provincial Air Service aircraft G-CAOS, had to be repaired and restored to display condition. Many other parts were made from drawings and/or badly corroded parts that had been salvaged, including the three main fuel tanks, the fuel gauges and the air-driven fuel pump and other fuel system components, all of which were completed to working condition.

The main structure was completed in the fall of 1984 and a trial assembly was made of all main components. The covering and doping which came next was a major job, not because of any unusual difficulty but because of the large area of the components and large number of hand stitches required. The final operations were the fitting of the flight controls, the engine and radiator controls, and the cockpit instruments and other details. Everything was completed by June 1986. It is now on public display in the new building.

In this view, taken on May 14, 1979, the hull is structurally complete, and cockpit openings have been cut and fitted with their coamings. A trial installation of the tail surfaces is also being carried out. (KMM)

McDowall Monoplane

The restoration of this diminutive machine began in the fall of 1983 and proceeded slowly, as work on the larger HS-2L restoration was also underway.

Work began with the wings and the Anzani engine. The engine, as usual, was torn down, cleaned and protected and reassembled.

The wooden parts required the splicing in of new pieces to replace those removed when in 1916 or 1917, it was intended to fit ailerons along with the usual minor repairs, cleaning and refinishing prior to covering. The tail surfaces required similar treatment. The fuselage required the usual disassembling of all parts, cleaning and refinishing, together with repairs as required. The cane bottom of the pilot's seat had to be replaced. The tires, of course, had to be replaced and the Museum was fortunate in obtaining new ones in the original white rubber.

A difficult decision was how to treat the fabric covering. The original linseed oil treatment was never very satisfactory for a number of reasons, but nevertheless was used extensively in the pre-World War I days. The use of dope would give a much more satisfactory finish but it would not look right historically. Experiments were carried out with tinted dope but the results were not deemed satisfactory, and the decision was taken to use linseed oil. It is believed to be the only museum specimen to be restored with this pioneer finish.

A detail of a typical fuselage joint in the McDowall Monoplane, the same type of joint as was used in the Blériot XI. (KMM)

A detail of the McDowall tail surfaces, covered but not yet treated with linseed oil, used in place of the modern dope finish. (KMM)

The Rolls-Royce Merlin installation in the Hispano HA-1112-M1L during its restoration.
(KMM)

A view of the cockpit of the Hispano HA-1112-M1L after its restoration.
(KMM)

Hispano HA-1112-M1L

A late decision was taken to restore this Spanish-built version of the Messerschmitt Bf 109 so it could be displayed in the new building. To expedite the work, the restoration was not as complete as usual and the engine was not stripped down. However, the fuselage was stripped of all equipment and everything was cleaned, repaired and reinstalled. The fuselage itself required some skin repairs, and missing access panels had to be replaced, but it was generally complete. The wings required some skin repairs and a new clear plastic navigation light cover, together with the usual cleaning and refinishing.

The aircraft was refinished in its original Spanish Air Force markings. The restoration was completed in early November 1986. A number of these aircraft have been restored in German markings to represent the World War II Messerschmitt Bf 109, but the inverted Vee Daimler-Benz engine of the German-built Bf 109 resulted in quite a marked difference in appearance to the Spanish-built machine with its Rolls-Royce Merlin engine, so a reasonable reproduction is not possible without re-engining the machine and other detail changes. This is a substantial job and in the majority of cases has not been done.

Threw the National Aviation Museum was fortunate to have Paul Hartman as Chief Pilot from the beginning of its flight operations until 1979. Indeed, he flew the Silver Dart reproduction even before the Museum was formed. Not only is Paul Hartman a fine pilot, he is an experienced test pilot and a graduate of the Empire Test Pilots School in 1948, and thus well qualified to assess the flying characteristics of the Museum's specimens.

He was born in Grafton, Massachusetts, in 1918 and, growing up in the golden age of aviation, became fascinated by aircraft and flying. He obtained his American private pilot's licence in 1938. Finding he was under the height limitation of the USAAC, he enlisted in the RCAF in 1941 and was awarded his wings later that year.

During World War II, he ferried a Lockheed Hudson overseas in April 1942, and served with No. 69 Squadron (RAF) flying Vickers Wellington Mk. VIIIs on night torpedo bomber operations. He then served as an instructor at No. 6 (Transport) OTU at Comox, British Columbia. He was awarded both the DFC and AFC for his wartime services.

Following the war, Hartman remained in the RCAF until his retirement in 1968 as a Wing Commander. He became a Canadian citizen in 1951, and served three terms as a test pilot at the RCAF's Central Experimental and Proving Establishment. Since his retirement he has continued his test flying as a freelance test pilot and to date has flown over 200 different types of aircraft.

It is believed that in flying the Museum's aircraft Paul Hartman has become, along with Cole Palen of Rhinebeck, New York, one of only two modern pilots to fly with the shoulder-yoke control system suggested by Dr. Alexander Graham Bell, a system adopted by the Aeronautical Experiment Association for their aircraft and standardized by Glenn Curtiss for his machines until 1914. Also he is believed to have handled more different types of rotary engines than any other modern pilot, including the Bentley B.R.2, 130 hp Clerget, and the 80 and 110 hp Le Rhônes.   K.M.M.

PAUL A. HARTMAN
(W/C, RCAF *retired*)

Flying the Museum's Aircraft

W/C Paul Hartman flying the Silver Dart *at Trenton, Ontario, in January 1959, with an experimental fin added to improve its directional control. Note the tufts fitted on the rudder to check the air flow.*

(TURBO TARLING VIA L. MILBERRY)

Flying the Silver Dart

I n the fall of 1957 the RCAF began the construction of a non-flying reproduction of the A.E.A. *Silver Dart*, under the direction of LAC Lionel McCaffery, to help celebrate the 50th anniversary of powered flight in Canada. Upon seeing the reproduction on Air Force Day at Trenton in 1958, the Chief of the Air Staff, A/M C.R. Slemon, decided that an airworthy *Silver Dart* reproduction should be made and flown at Baddeck, Nova Scotia, on the anniversary.

A technical team under F/L W.K. Bell was set up to make the airworthy reproduction. A check showed only slight strengthening was necessary to make a satisfactory airframe. An attempt to have a reproduction of the original Curtiss engine made by the Curtiss-Wright Corp. was dropped as being too costly and a modern Continental A-65 of 65 hp was substituted.

248

Although the original *Silver Dart* had flown successfully in 1908 and 1909, few if any figures were available on actual flight characteristics, control forces and the control positions at take-off and in level flight; nor was information on its longitudinal stability available. To help in predicting flight characteristics a six-foot wing section was mounted on a truck, which was driven along the Trenton runway while measurements were made of lift, drag and the pitching moments of the wing. Tests were also made to determine the elevator angle required for take-off and it was decided to fit a simple wind-vane type of airspeed indicator, which, of course, the original did not have.

In the meantime, I had seen the aircraft and remarked to an associate that it looked like an interesting machine and I wondered what it would be like to fly. I was soon to find out, for I received a note from the Chief of Air Staff saying that I had been selected to fly the *Silver Dart*.

The aircraft was rolled out for its initial flight on January 29, 1959. First, two short runs were made up to 30 mph to check the aileron effectiveness. The ailerons appeared effective, although they produced a very low rate of roll of about 3-5 degrees per second. This was not surprising considering the high aspect ratio wing and small aileron area.

The first flight was made by setting the elevator at 5° incidence and opening the throttle to give 50 hp. The aircraft was easy to keep straight and lifted clear of the runway at 30 mph. It rose to a height of about 6 to 8 inches and no pitch up or down was noted. On closing the throttle the aircraft settled on the ground after a flight of 150 to 200 feet.

All flights were short flights or hops in a straight line. Strict instructions had been issued by Air Force Headquarters on how and to what extent the *Silver Dart* was to be flown: no turns were permitted and its altitude was limited to 20 feet! The directive made no allowance for "acts of God" and the limitations were exceeded occasionally.

The second flight followed immediately upon returning to the starting point and it covered about 500 feet at 10 or 12 feet altitude. On reducing power the machine settled on the ground in a three-point landing at 30 mph. During this flight the front elevator was moved each side of the 5° take-off position. It proved very effective and small changes in elevator angle produced marked changes in aircraft pitch. The lack of friction in the control system, combined with a lack of aerodynamic forces, produced somewhat disconcerting longitudinal control characteristics as there was no "feel" to help the pilot. This tended to produce an oscillating flight path, a feature that was also apparent in other early aircraft.

On the third flight we learned a good deal about the aircraft's

W/C Hartman flying the Silver Dart *at Baddeck, N.S., on February 21, 1959, the day before the 50th anniversary of the original flight of the first* Silver Dart.

(DND PL 114532)

directional control and stability. At an altitude of about 20 feet, the *Silver Dart* drifted to the right because of a light crosswind. About half rudder and aileron were applied, and after the heading had changed about 5°, the controls were centralized. However, the aircraft continued to slip to the right, travelling along its original flight path, and the nose began a slow swing to the left. Full right rudder had no effect in correcting the swing and it appeared that the rudder stalled when moved beyond its halfway-over position. The sideslip was overcome by banking 5-6° to the left and lowering the nose.

After regaining the original flight path over the runway, I reduced the power to idle and attempted a full-stall, power-off landing. The landing flare was started at 30 mph, about 5 feet off the ground with the nose up about 9-10°. The aircraft ceased flying abruptly at 27 mph, and the left wing dropped and struck the ground. There was no warning – it just stopped flying. However, the front elevator did not stall, which was fortunate, for if the nose had dropped the pilot might have been skewered by the push-pull elevator control rod.

The lower left wing tip was damaged and the lower bamboo structure at the rear collapsed. The flight was very educational. We now knew that the *Dart* must not be allowed to deviate from straight flight and that power-on landings must be made. The lack of directional stability was disconcerting, especially when combined with a rudder that stalled when deflected beyond the halfway point.

The aircraft was repaired and a triangular vertical fin 12 square feet in area was installed ahead of the rudder. The machine now had good directional stability but, for the sake of authenticity, we had to revert to the original configuration for the Baddeck flights. After further flights on February 5, the *Dart* was loaded in an RCAF Fairchild C-119 Boxcar and flown to Sydney, Nova Scotia. It was then trucked to Baddeck, where it was assembled in a polyethylene-covered, wood-frame hangar erected for the occasion on the north shore of the ice-covered Great Bras d'Or.

On February 21, the *Silver Dart* was flown for the ninth time, making the best flight of its total of ten flights. The weather was clear with no wind; a distance of just over a half mile was covered, and a maximum speed of 43 mph, the highest ever, was achieved. At that speed the aircraft was becoming longitudinally unstable and I had to apply nose-up elevator to prevent the *Dart* from pitching down above 40 mph.

While I was unbuckling my seat belt after landing, a tall elderly gentleman came over from an RCAF DHC-3 Otter that had landed just before I took off. He shook my hand and thanked me for the pleasure I had given him by enabling him to see the *Silver Dart* in flight for the first time. This was my first meeting with the Honourable J.A.D. McCurdy.*

* McCurdy had, apparently, forgotten that F.W. (Casey) Baldwin had made five flights in the original *Silver Dart* at Baddeck when he was present. K.M.M.

250

The anniversary date, February 23, dawned clear, cold and calm, but as the scheduled flight time approached a west wind arose to whip up the flags and bunting and chill the spectators. The wind caused concern, as the Silver Dart was difficult, if not impossible, to control in winds over 10 mph. But the wind then dropped and a decision was made to fly. The Silver Dart took off at 11:30.

Immediately the wind increased with gusts of 15 to 20 mph tossing the aircraft up to about 100 feet before I could regain control. Full throttle was required to enable the Dart to move forward against the wind, which appeared to be gusting to about 30-35 mph. The aircraft was almost out of control and the ailerons and rudder were completely ineffective; however the elevator was still effective. The flight became analogous to trying to control a cardboard box in a high wind. I landed by closing the throttle and allowing the aircraft to stall on to the ice. The left wing dropped at the stall about 15 feet above the ice. The final impact was cushioned by the wing and it felt no worse than a heavy landing.

The flight time accumulated on the Silver Dart was about four minutes during its ten flights. Flying this aircraft was an interesting and informative experience but one which I have no desire to repeat.

Sopwith Snipe

The next Museum aircraft I flew was the Sopwith Snipe obtained in the United States in the winter of 1964. It was decided to fly it at the Air Force Day celebration at Rockcliffe on June 6, 1964, honouring the RCAF's 40th anniversary, and I had the good fortune to be selected as the pilot.

The Snipe was the last RAF fighter to go into service in World War I, and although its performance had proved disappointing, it looked every inch a real fighter to me. I was impressed. It would be the first World War I fighter I had flown and the first rotary-engined machine as well. It had quite a complete set of instruments compared with other Museum aircraft that I later flew, including an airspeed indicator, altimeter, inclinometer, compass, tachometer and pulsometer. The engine controls consisting of a throttle, a fine fuel adjustment lever and the pulsometer were typical of the rotary-engined machines.

The rotary engine revolved completely, unlike the more usual static radial engine, so the propeller was attached to the crankcase and the crankshaft to the airframe. The rotary used castor oil, which was insoluble in gasoline, as a lubricant. The oil entered the engine with the gasoline, which was supplied through the crankshaft, and the globules of castor oil, which varied in quantity with engine speed, could be seen in the pulsometer, providing a constant check on the lubrication of the engine. The smell of burnt castor oil drifting downwind from

rotary-engined aircraft was an unforgettable part of the early aviation scene, an experience only available now to those who have the rare privilege of seeing and smelling a rotary-powered aircraft during a flight demonstration.

The gyroscopic effect of the rotary engine affected the flight characteristics of the aircraft, but how much they were affected depended on the size and design of the particular type. Of the fighters I flew, the Camel was the most affected and the Snipe the least. The rotary engine would cause the nose to drop in a right turn, which could lead to a spin if not corrected, and in a left turn the nose would rise.

My first flight in the Snipe took place on May 21, 1964, and in flying it I found the Bentley B.R.2 engine more sensitive to the fine-adjustment fuel mixture setting than the other rotaries. On take-off at full throttle at 1,250 rpm, the tail left the ground at about 30 knots and a slight tendency to swing to the left was easily corrected. At 45 knots a slight movement of the stick aft caused the machine to become airborne; it accelerated to 65 knots and then held in steady climb.

Its best climbing speed was 65 knots but the lower nose position with better visibility at 70 knots proved to be more comfortable. The Snipe's excellent stability about all three axes served to mask the normal gyroscopic effect when manoeuvring. The aircraft was flown down to 41 knots with engine idling, and the rudder and ailerons were effective in maintaining control, but I did not put it into a complete power-off stall.

A level speed run at 1,150 rpm gave a true airspeed of 98 knots at 1,500 feet, which agreed with official World War I figures and closely corresponded with its predecessor, the 1F.1 Camel with only 130 hp. No attempt was made to dive the Snipe or indulge in aerobatics, so a maximum speed of 130 knots indicated airspeed recorded in a slight dive was the highest attained. A gliding approach at 65 knots was maintained by blipping the engine (i.e. cutting out the engine by means of a switch on the control stick) and the speed was reduced progressively to allow the landing flare to begin at 42-43 knots about five feet above the ground. The speed decreased to 40 knots and the aircraft attained a tail-down position about a foot off the ground before making contact in the usual three-point attitude. The steerable tail skid made the Snipe (and the similarly equipped Camel) the easiest of the World War I machines to keep straight during the roll following landing. However, the tail skid did not permit making turns of 90° or more on the ground; such turns had to be made by easing the stick forward, opening throttle and applying rudder. The elevator and rudder controls were light and effective, but the aileron control was heavier and not as pleasant as on the other Sopwith types I was to fly later.

Nieuport 17

The next World War I aircraft type I flew was a reproduction Nieuport 17 made to original drawings and powered with an original 110 hp Le Rhône rotary engine. The type first appeared in early 1916 but was in fact merely a slightly enlarged and more powerful version of the Nieuport 11, the "Bébé" Nieuport of 1915.

The cockpit instrumentation was sparse, consisting of an inclinometer, tachometer and pulsometer. We added an airspeed indicator for our flying. The usual rotary engine controls and flight controls were fitted.

My first flight in the Nieuport 17 took place on May 4, 1967. The engine caught on the first swing of the propeller and, after an alteration of the fine adjustment setting, idled smoothly at 750 rpm. A quick check of the controls verified that they were working correctly. A glance at the pulsometer showed satisfactory lubrication, but soon I learned that a film of oil building up on the windscreen and leading edges of the lower wings was the best check of adequate rotary engine lubrication.

The wheel chocks were removed and I opened the throttle slowly. When the rpm reached 1,100 and the machine had travelled about 150 feet, I began to move the stick forward, but before I could complete the

movement the aircraft left the ground in a three-point attitude at about 38 knots. Once airborne, the left wing dropped slightly and the nose swung to the left, but both were easily corrected with rudder and ailerons. The rudder was light and quite powerful and the aircraft responded to it excellently. It also responded well to the ailerons but they were surprisingly heavy to operate.

I held the aircraft down to accelerate to 60 knots and started to climb. As the climbing angle was so steep that forward visibility was restricted, I increased the speed to 70 knots. At approximately 1,000 feet – no altimeter was fitted – I levelled off and started a left turn. The gyroscopic effect of the rotary engine was immediately felt and corrected with rudder.

Following the turn, I continued the climb to about 1,500 feet, where I made a maximum level speed run. At 1,200 rpm the speed stabilized at 80 knots indicated airspeed, later estimated as 98 mph true airspeed, slightly under its recorded speed of 102 mph in World War I. At the increased speed it was necessary to push the stick forward with a force of about 25-30 lb to maintain level flight and to counteract with rudder a swing to the left. I then made a shallow dive until 105 knots was reached, which required about a 50-lb push on the stick to keep the machine in the dive. The aircraft was stable longitudinally and laterally in the dive but right rudder was required to correct the yaw. I found the observed action of the air bubble in the inclinometer quite disconcerting as it reacted differently than the metal ball in modern bank indicators.

After the dive, I tried a stall at approximately 1,000 feet. The throttle was closed and the stick was slowly pulled back. As the speed decreased to 43 knots, the airspeed needle oscillated about plus or minus 2 knots. A light tail buffet started at about 40 knots with the stick fully back and engine idling at 700 rpm. With the nose held high, the speed dropped to 38 knots and stabilized, and I realized that the idling rpm prevented stalling. So I depressed the blip switch, the rpm dropped to 400, the speed fell off to 33-35 knots, and the left wing dropped sharply. The ailerons proved ineffective in raising the wing and seemed to accelerate the rate of wing dropping. The nose did not drop at the stall. When I eased the stick forward and released the blip switch, the engine responded, and the aircraft recovered from the stall with only a reasonable loss of altitude.

A glide was then established at 50 knots. The rate of descent was not excessive and the aircraft responded well to the controls, but its nose-high attitude was uncomfortable as it restricted forward visibility. A steeper glide was then maintained by blipping the engine to maintain 50 knots with the nose held down.

The Nieuport 17 proved easy to land. I held the glide until about ten feet from the ground and started the flare as the landing area was reached. About two or three feet off the ground, I held the blip switch down and eased the stick fully back. The tail dropped and I made a reasonably decent three-point landing. There was no difficulty in keeping the aircraft heading straight during the run-out.

We were concerned about the heavy stick force necessary to maintain level flight at 80 knots. The horizontal stabilizer was not adjustable either in the air or on the ground, and while we might have experimented by adding a short length of cord to the elevator trailing edge, it was felt that this might lead to other unacceptable flight characteristics. We then realized that altering the incidence of the lower wings might help and increased the angle by 2° to 5°, the maximum allowable. This reduced the stick force to 15 lb. The previous 25-30 lb, although acceptable for short flights, would become tiring after about 20 minutes. We then added a 10-lb spring to the front of the control stick at about eight inches above its pivot point. These adjustments resulted in the Nieuport being very pleasant to fly at speeds up to 105 knots, the maximum intended for Museum use.

The heavy aileron stick forces, together with the light forces required for the elevator and rudder, were the only disappointing features of the Nieuport, although I am not sure my assessment is entirely fair as it may be that I tend to compare its handling characteristics too closely with modern fighters. As mentioned elsewhere, the Nieuport was damaged in an accident before the Air Force Day display in 1967. It was subsequently repaired and has appeared at various Canadian centres, and I have flown it on several of these occasions.

Sopwith Triplane

On May 5, 1967, the day after flying the Nieuport 17 for the first time, I flew the Sopwith Triplane. This was a reproduction aircraft like the Nieuport and powered with an original 130 hp Clerget engine. Of all the World War I aircraft I flew, this was undoubtedly the most pleasant. It possessed stability and control response so good as to be unique in comparison with its contemporaries and also compared very favourably with modern machines. Its positive stability about all three axes, together with a stabilizer that was adjustable in flight, resulted in an aircraft that could be flown hands-off. The controls were light, powerful and well harmonized, and the aircraft responded to them excellently.

The engine controls were the same as for other rotary engines and the aircraft was equipped with both altimeter and airspeed indicator. The pilot was seated lower than in the other machines, giving an initial impression of limited forward visibility. However, a gap between the roots of the centre mainplane and the fuselage sides enabled the pilot to see forward and downward for the approach and landing. The single Vickers gun was removed as a precaution against any difficulty that might be experienced in my first landing.

The engine started easily, as did all the Museum's rotaries, and idled smoothly, but the French tachometer installed gave slightly erratic readings of twice engine speed. As I was determined not to allow the Triplane to become airborne in the three-point position as the

Nieuport had done the day before, I held the stick slightly forward as I opened the throttle. The aircraft accelerated quickly and the tail rose at about 20 knots. It had no tendency to swing and very little rudder was required to keep it straight on the take-off run. With the tail up, visibility over the nose was excellent.

The Triplane left the ground at 38-40 knots and rapidly accelerated to 60 knots, and I continued to climb at 60. The altimeter read 650-700 feet as I reached the airport boundary, about 5,000 feet from the take-off point. A ground observer later said that the Triplane appeared to be climbing like the proverbial "homesick angel" and it certainly felt that way to me.

I levelled off at 900 feet and commenced a 40° turn to the left. There was no tendency to slip or skid, and no gyroscopic effect from the engine was felt. At 1,500 feet I checked the stall. I closed the throttle the engine idled at 700 rpm and a speed of 40 knots was indicated. The nose was held high at about 12-13° and refused to stall. When I depressed the blip switch, the rpm decreased and the nose dropped. The ailerons remained fully effective throughout the stall and there was no tendency for a wing to drop. As the nose dropped, I released the blip switch and the engine idled. There was a height loss of about 75 feet during the recovery.

Next the glide was checked and the aircraft was trimmed to glide at 50 knots. Generous use of the blip switch was made during the glide, as the machine would not descend at an acceptable rate unless the rpm was reduced below idling. I took care not to cut the ignition for more than four or five seconds at a time as a World War I manual noted that prolonged use of the blip switch without shutting off the fuel supply could lead to an accumulation of a combustible fuel/air mixture within the engine cowling. This mixture, it was said, could produce surprising and spectacular results when the ignition was turned on again.

The glide was continued to 1,000 feet altitude and a level speed run was carried out. After correction it was found a true airspeed of 111 mph was attained, which agreed almost exactly with the official World War I test figures. I then checked the effectiveness of the stabilizer trimming arrangement and found it to be a positive and powerful device. The maximum speed at which the aircraft was flown was 105 knots in a shallow dive. It had a slight tendency to yaw to the left as speed increased and to yaw to the right as it was reduced, which was easily corrected with the rudder.

Landing was easy. I glided the aircraft to about 15 feet from the ground and pressed the blip switch. As I eased the stick back, the machine slowly lost speed until it was about a foot off the ground at 40 knots. The stick was then brought fully back and the aircraft was on the ground at 37-38 knots.

Sopwith 2F.1 Camel

On May 26, 1967, I made my first flight in the Camel, which was an original machine overhauled and restored. The Museum's Camel was the 2F.1 Camel, the naval version of the more numerous 1F.1 Camel, and it had a wing span about a foot shorter. With its reputation and fine combat record, I looked forward to my first flight with much interest but at the same time some apprehension. In some quarters it was regarded as a fierce little beast, a pilot killer and a "popping little firecracker." This apprehension was not reduced when, as I was sitting in the cockpit before my initial flight, an elderly gentleman approached, introduced himself as a former Camel pilot and launched into an account of the number of pilots killed in their first Camel flight.

I pressurized the fuel tank by hand pump to 2-3 psi; the Clerget engine started easily and I switched over to the engine driven air pump. Incidentally, all flights of these early machines were conducted from the grass surface of the airfield, not the paved runways, and the steerable tail skid of the Camel helped to overcome a slight swing to the left as I opened the throttle. The tail came up at about 20 knots.

As the speed reached 40 knots I eased the stick back slightly and the Camel became airborne. It accelerated rapidly to 55 knots and I held it in a climb at that speed until I reached 500 feet. I levelled off, leaned the mixture and started a left turn. The gyroscopic effect from the rotary

258

engine quickly became apparent. As the bank angle approached 40° and I applied a slight amount of elevator, the nose rose noticeably and the machine slipped to the left. An immediate application of bottom rudder to depress the nose tightened the turn. By now the bank angle was about 50° and I eased the stick forward to prevent the turn tightening. This resulted in the nose dropping slightly. I then found that the right turn was also affected, with the aircraft yawing to the right and the nose dropping when entering the turn. Applying left rudder to raise the nose tightened the turn, and a forward motion of the stick induced a side slip to the right.

The Camel was not difficult to fly in turns but merely different from the other machines. The gyroscopic effects were more pronounced but were easily overcome by moving the controls smoothly and not too rapidly when initiating manoeuvres. The elevators and rudder were light and powerful and the aircraft responded quickly to them. As I became more familiar with the Camel, I realized an especially light touch was needed on the stick to feel the forces acting on it. In precise co-ordinated turns, control was best effected by applying barely perceptible movements to the stick with very light touches on the rudder. The Camel responded so rapidly to small rudder movements that it was difficult not to over-control it in yaw. I finally resorted to wearing thin-soled shoes for a better feel of the rudder forces, and this made an appreciable difference.

It became apparent in steep turns that the aircraft lacked dynamic longitudinal stability, as it tended to pitch up, tightening the turn as speed decreased with the increased lift and drag. This was observed initially by the control inputs required to overcome the gyroscopic effects encountered on entry to the turn, but it was also evident as the turn was stabilized. Constant forward pressure on the stick was required to combat the effect, which varied with engine rpm and airspeed. Increasing the rpm with constant speed reduced the pressure required, while a decrease in speed with constant rpm, increased pressure needed. This instability was also apparent in the aircraft behaviour as it approached the stall.

The stall checks were made at 2,500 feet and the Camel's characteristics in the approach to the stall were similar to, but more pronounced than, those of the Snipe. As the speed decreased to 40-41 knots with engine idling, the pull force required on the stick decreased appreciably. Full aft movement of the stick reduced the speed to 38-39 knots and the aircraft assumed a stabilized nose-high altitude. Practically no force was required to hold the Camel there, and the machine was still responsive to aileron and rudder control. I pressed the blip switch and the speed dropped to 36-37 knots. Light tail buffeting began, the tail dropped, and the nose pitched up. Release of the blip switch and a rapid movement of the stick fully forward caused the nose to drop. The height loss during the stall recovery was about 75 feet.

The gliding characteristics were then checked at 50 knots and with the engine idling at 700 rpm. The stability and control were satisfactory but the rate of descent was unacceptably low, so it was increased to about 500 feet per minute by intermittent use of the blip switch. The response to the controls in gliding turns in both directions was excellent and the gyroscopic effect from the idling engine almost indiscernible. Left and right side slips in both turning and straight flight were made, and the aircraft responded rapidly and precisely.

The maximum speed of 97 knots attained in a level speed run at 1,000 feet agreed with recorded World War I figures. Fairly pronounced changes in pitch and yaw occurred with increasing speed, and the nose tended to rise and swing to the left. These trim changes induced gyroscopic moments from the engine that accentuated the nose-up pitch but tended to counter the left yaw. Forward movement of the stick to lower the nose set up an additional gyroscopic moment and increased the left yaw, and a correction by right rudder induced more nose-down pitch. Although these trim changes were quite apparent, they were easily corrected with slight aileron and rudder movements. The result of this confusing situation was that I held the machine steady during the speed run by holding the rudder fixed and applying short, jabbing movements to the stick. The resulting stick movement was almost circular and resembled "a slight stirring of the pot."

The Camel appeared to have static longitudinal stability. The push force on the stick to maintain altitude increased with the speed and about 15 lb had to be applied at 97 knots. The Camel did not have an adjustable horizontal stabilizer like the Triplane and Snipe. A later ground adjustment of the stabilizer setting of about 1° reduced the required stick force at 97 knots to about 10 lb. This setting change did not affect the response in turns.

The maximum speed reached in a shallow dive at 3,000 feet was 130 knots. The trim changes and gyroscopic effects noted in the speed runs were again observed but were more pronounced. Any slight, abrupt forward movement of the stick both steepened the dive and caused the nose to veer left, and corrective right rudder induced nose-down pitch. Therefore the stick was handled carefully in dives.

The instability at the stall made the Camel easy to land, like the Snipe. The glide at 50 knots was held to about 20 feet from the ground. Then the blip switch was held, speed decreased to 45 knots and the landing flare started about eight to ten feet off the ground. Speed and height were lost until the aircraft was level at about three feet off the ground at 40-41 knots. The stick was moved aft as soon as the sink began, and the machine settled on the ground in a three-point attitude.

I enjoyed flying the Camel, but its vices of control instability, extreme control sensitivity and pronounced gyroscopic effects all combined to create the impression of balancing an egg on the point of a needle rather than flying an aircraft. It was a highly manoeuvrable machine that was best flown by exploiting its deficiencies with a firm but light touch on the controls. It was never forced into manoeuvres – they were executed by light pressure on the controls and subsequently relaxing, or sometimes reversing, the pressure once the desired rate of response was attained.

By modern standards of stability and control, the Camel would be totally unacceptable as a military aircraft. Its unorthodox handling qualities were the price of its extreme manoeuvrability, which enabled its pilots to shoot down more enemy aircraft than those of any other British type. Its flying qualities were such that pilots either loved it or despised it, but none treated it with indifference.

F/L J. McKay flying the Fleet 16B Finch II in June 1967.
(DND PC 67-12)

Fleet Finch and Aeronca C-2

The next two Museum aircraft I flew were products of a later era and not in the same class as the World War I machines. Their flying qualities closely approximated those of the better light aircraft of today. The first of these was the Fleet 16B Finch II, a World War II trainer that was, in fact, the final development to RCAF requirements of a civil trainer designed by Consolidated Aircraft Corp. in 1928. I first flew the

Fleet on April 25, 1967. I found it stable about all three axes and quite manoeuvrable, and it responded rapidly to its light, well harmonized controls. It was a pleasant aircraft to fly.

It took off at about 55 mph, climbed well at 65 mph and attained a maximum speed of level flight of 135 mph when tested without its usual World War II crew enclosure installed. While I did not attempt aerobatics in the Finch, its similar predecessor in the 1930s established a remarkable reputation in the RCAF for its strength, and pilots stunted it with great confidence. Like most aircraft, it also possessed one poor characteristic. Its narrow undercarriage and long-stroke shock absorbers made it tricky to land in a cross wind. During my first landing at Rockcliffe, I left a fluff of Finch wing feather on the runway immediately after touching down. The aircraft was undamaged, although I could not say the same for my pride.

I flew the Aeronca C-2 for the first time on June 9, 1967. It was a low-powered machine designed to be sold cheaply to private pilots and to be economical to use. Its design began in 1925 but it only entered production in 1929, and the Museum's specimen was only the eighth made that year. The specimen had a slight warp in the right wing which precluded an assessment of its lateral stability, as aileron deflection was required to maintain level flight at all times.

It took off at 40-41 mph and climbed best at 55 mph. Its climb performance was poor, and a later check showed it was losing compression in one cylinder. In spite of the power loss, a maximum speed of 80 mph was attained in level flight. The Aeronca appeared to be an efficient low-powered machine, as its pilot for the Air Force Day display weighed 210 lb, which it handled even with its engine operating at reduced power.

The Museum's Avro 504K G-CYCK climbing out at Rockcliffe, June 4, 1972, flown by Paul Hartman. (KMM)

Avro 504K

The next Museum type I flew was another World War I machine, the Avro 504K, and I took it into the air for the first time at Rockcliffe on June 5, 1967. It was the largest and, since its design originated in 1913, the oldest of all the Museum's World War I machines I had flown. Avro 504Ks were fitted with several different engines. The machine I flew, G-CYCK, was fitted with the 110 Le Rhône rotary engine. Another Museum 504K, 'FG, is fitted with the 130 hp Clerget and is usually on static display. Despite being somewhat underpowered, the 504 was surprisingly manoeuvrable. It displayed no vices and the elevator, aileron and rudder controls were fully effective in controlling it during all flight manoeuvres. It displayed no tendency to spin at the stall and its long fuselage and large-span wings effectively damped out the

gyroscopic moments arising from the rotary engine during manoeuvres.

The landing gear skid, whose curved front end projected about two feet forward of the propeller, was the only inherently bad design feature of the aircraft. Apparently intended to prevent the propeller striking the ground during a tail-high take-off or landing, experience showed that if the skid struck the ground the aircraft almost always flipped onto its back. Thus the skid acted as inducement to the making of proper landings! Fortunately, the 504 was a very easy aircraft in which to make the classic three-point landing.

During the take-off, which was made with the tail kept slightly below the level flight attitude, a touch of right rudder sufficed to keep the aircraft straight during the initial acceleration to lift-off at approximately 40 mph. With engine at full power, 1,200 rpm, the aircraft accelerated quickly to its best climbing speed, 55 mph. A rate of climb of 500 feet per minute was obtained to 2,000 feet, where a maximum indicated air speed of 83 mph or 86 mph true airspeed was achieved in level flight, well below the officially recorded World War I speed of 95 mph at sea level.

In the approach to the stall with the engine idling at 650 rpm, the controls became sloppy and ineffective as the speed decreased to 30 mph. With the control stick fully back at 25 mph, the left wing dropped and was followed by the nose. The stall was quite gentle and the recovery, which was made by centring the controls and increasing the power to about 1,000 rpm, was achieved for a loss of less than 100 feet in altitude.

The glide to a landing was made at 55 mph, its best gliding speed, and the desired rate of descent was held at 650 feet per minute by the intermittent use of the blip switch, for, as on the other World War I machines, the large propeller idling at 650 rpm gave considerable thrust. Within about 50 feet of the ground, the blip switch was held depressed as the machine was slowly flared to level flight two or three feet above the ground. As the airspeed decreased and the aircraft started to sink, the stick was moved fully aft and the aircraft settled onto the ground in a three-point attitude. In winds of 5-10 mph the aircraft rolled to a stop in less than 50 yards.

As it was a Museum aircraft, aerobatics were prohibited, but all normal manoeuvres, including steep turns to 70-80° of bank, sideslips and stalls, were performed and a maximum speed of 110 mph was attained in a shallow dive. The aircraft responded well to all controls and its general handling characteristics were good. Assessed against modern standards, however, the control harmonization (or ratios of lateral, longitudinal and directional control stick and rudder bar movements in executing manoeuvres requiring precise co-ordination of controls) was poor. A disproportionately large amount of lateral displacement of the stick was necessary to achieve a desired rate of roll relative to the amount of rudder or elevator required for directional or longitudinal control. Despite this shortcoming, the aircraft was easy and pleasant to fly, and undoubtedly it was these qualities that were responsible for the 504's long use as a pilot trainer in many countries throughout the world.

Sopwith Pup

Occasionally a pilot will be lucky enough to fly an aircraft whose delightful flying qualities make it a mere extension of himself, so that he feels it is "strapped to his backside." The Sopwith Pup of 1916, a beautifully proportioned little biplane with an 80 hp Le Rhône rotary engine, was such an aircraft and it was my good fortune to fly it first at Rockcliffe on August 10, 1974.

The Pup was almost as agile as the Camel but possessed none of its vices. Its handling qualities were excellent and the controls were light and well harmonized and it responded immediately to a slight displacement of the controls without being overly sensitive. At its best

climbing speed of 50 mph, with the engine delivering 1,100 rpm, an initial rate of climb of 700 feet per minute was obtained. In level flight at 2,000 feet and at 1,100 rpm, a maximum indicated airspeed of 98 mph (102 mph true airspeed) was attained, just under its officially recorded World War I speed.

The power-off stall occurred at about 35-36 mph and was manifest as a gentle nose-down pitch. Recovery from the stall was immediate following release of the blip switch and centring the stick. The loss in height during the stall did not exceed 75 feet. The aircraft appeared to glide best at about 60 mph with the engine idling at 650 rpm. At that airspeed, the nose was sufficiently below the horizon to provide adequate forward visibility with a rate of descent of approximately 500 feet per minute. During the landing approach, I depressed the blip switch periodically to prevent the speed increasing. About 50 feet above the ground the aircraft was slowly rounded out to level flight at two or three feet above the ground and the blip switch was held depressed. Height was maintained as the speed dropped by slowly moving the stick rearward. When the stick was fully aft and the tail well down, the aircraft started to settle onto the ground in a three-point landing. The landing roll-out was less than 50 yards.

Flying the Pup was a most pleasant experience and I concur fully with that majority of ex-RFC pilots who unhesitatingly cited it as the supreme example of an aircraft with the most delightful flying qualities and devoid of all vices.

As mentioned in the history section of this book, a large number of German aircraft were brought to Canada in 1919 as part of a war trophy collection acquired by Lt. Col. Arthur Doughty (later Sir Arthur) for the Canadian government. Unfortunately only three of these aircraft are known to have survived. Of these, the A.E.G. G.IV and Junkers J.1 are unique specimens and are in the Museum's collection, for which Canadians can be grateful. The complete known list of these aircraft is presented here with serial numbers and notes on their fate. Where their fate is not known it seems almost certain they were among those destroyed at Camp Borden, apparently in 1921.

Aviatik

Type and serial number not recorded, "No.X G.13." Via SS *Commonwealth* June 15, 1919. On loan from RAF.

A.E.G. G.IV

574/18/ From Dieppe via SS *Venusia* May 23, 1919. At Toronto 1919. Much later to Canadian War Museum for storage. Less engines and nacelles. Restored by 6 RD, RCAF, 1968-69. On display at Rockcliffe.

Albatros D.V

2360/17. From Dieppe via SS *Venusia* May 23, 1919.

4698/17. From Dieppe via SS *Venusia* May 23, 1919.

Albatros D.Va

5821/17. Via SS *Montezuma* July 7, 1919. While recorded as a D.V, German records indicate that it was, in fact, a D.Va.

6530/17. From Dieppe via SS *Venusia* May 23, 1919.

Fokker D. VII 6810/18 (Alb.) now at Knowlton, Québec, in the Paul Holland Knowlton Museum of the Brome County Historical Society. Photographed after fabric repairs by RCAF in 1963, it is the only D. VII extant with original German dyed fabric. (KMM/H. TATE)

Fokker D.VII

5492/18 (Alb.). From Chingford via SS *Venusia* May 5, 1919.

6769/18 (Alb.). From Chingford via SS *Venusia* May 23, 1919.

6810/18 (Alb.). No shipping record from overseas or Camp Borden but it seems likely it was collected by Lt.-Col. Doughty. In Paul Holland Knowlton Museum of Brome County Historical Society, Knowlton, Quebec. Fabric repairs by 6 RD RCAF in 1963.

6822/18 (Alb.). From Chingford via SS *Venusia* May 5, 1919.

6832/18 (Alb.). Likely with No.1 Squad., CAF, Shoreham. Via SS *Peridat* June 24, 1919. To University of Saskatchewan, Saskatoon, from Camp Borden, May 7, 1920.

6933/18 (Alb.). From Chingford via SS *Venusia* May 23, 1919.

6842/18 (Alb.). No shipping record from overseas. To University of Manitoba, Winnipeg, from Camp Borden, May 7, 1920. Airframe demolished and engine retained in Engineering Faculty. Engine to NAM in an exchange in May 1975 and now installed in the Museum's Fokker D.VII 10347/18.

6846/18 (Alb.). From Chingford via SS *Venusia* May 5, 1919.

6849/18 (Alb.). From Chingford via SS *Montezuma* July 10, 1919.

7728/18. Fokker built. Via SS *Montezuma* July 10, 1919.

8413/18 (O.A.W.). From Dieppe via SS *Venusia* May 23, 1919.

8474/18 (O.A.W.). From Dieppe via SS *Venusia* May 23, 1919. Displayed at Canadian National Exhibition, Toronto, August 1919.

8488/18 (O.A.W.). From Chingford via SS *Venusia* May 23, 1919.

8492/18 (O.A.W.). From Chingford via SS *Venusia* May 23, 1919.

8493/18 (O.A.W.). With No.1 Squadron, CAF Shoreham. No shipping record from overseas. To University of Calgary from Camp Borden May 12, 1920. Displayed there July 6-8, 1920. Later disassembled and parts dispersed.

8502/18 (O.A.W.). From Dieppe via SS *Venusia* May 23, 1919.

8526/18 (O.A.W.). From Dieppe via SS *Venusia* May 23, 1919.

8583/18 (O.A.W.). From Dieppe via SS *Venusia* May 23, 1919. Assembled at Leaside Aerodrome and likely flown there August 1919. To McGill University, Montreal, from Camp Borden May 14, 1920. University has no record.

8609/18 (O.A.W.). From Dieppe via SS *Venusia* May 23, 1919.

10349/18. Fokker built. From Dieppe via SS *Venusia* May 23, 1919. Assembled at Leaside Aerodrome. Capt. L.B. Hyde-Pearson flew this aircraft through a tree in the fall

of 1918 and returned safely to Leaside Aerodrome, Toronto.

10350/18. Fokker built. From Dieppe via SS *Venusia* May 23, 1919. Assembled and likely flown at Leaside Aerodrome, Toronto, August 1919.

Fokker D. VII 8493/18 (O.A.W.) of No. 1 Squadron, CAF, Shoreham, England, 1919. Top scoring two-seater pilot, Canadian Capt. A.E. McKeever is in the cockpit. Fokker D. VIIs formed part of the equipment of No. 1 Squadron, CAF, whose emblem is seen here. This machine and others from the squadron came to Canada as war trophies. (PAC M816-N)

This Fokker D. VII 10349/18 was flown through a tree near Leaside Aerodrome, Toronto, in the fall of 1919, by Capt. L.B. Hyde-Pearson. The advertised Air Force Album is probably Alan Sullivan's Aviation in Canada 1917-18. *The Museum's D. VII is 10347/18, only two machines earlier than this one.* (CWM/A.G. McLERIE COLLECTION)

Fokker E.V. (D.VIII)

132/18. With CAF Shoreham. Shipped via SS *Peridat* June 24, 1919. Assembled, and likely flown, at Leaside Aerodrome, Toronto, August 1919.

Fokker E. V 132/18 at Shoreham, England, 1919, with Canadian Capt. G.O. Johnson, MC, CdeG, in the cockpit. It was later flown at Leaside Aerodrome, Toronto.
(A.E. de M. JARVIS)

Halberstadt C. V (Av.) 6863/18 at Shoreham, England, in 1919. (A.E. de M. JARVIS)

Halberstadt C.V(Av)

6863/18. From Chingford. Via SS *Montezuma* July 10, 1919.

Junkers J.I

586/18. From Dieppe via SS *Venusia* May 23, 1919. Displayed in damaged condition at the Canadian National Exhibition, Toronto, August 1919. Later stored by Cana-dian War Museum. Now in NAM at Rockcliffe. Has been displayed unrestored at Rockcliffe.

L.F.G. Roland D.VIb

6142/18. From Dieppe via SS *Venusia* May 23, 1919.

6145/18. From Dieppe via SS *Venusia* May 23, 1919. To Mayor of Winnipeg, from Camp Borden on August 12, 1920. On loan to Canadian Aircraft Co., St. Charles Aerodrome, Winnipeg. Nosed over in windstorm there in fall of 1920. Later acquired by W.P.A. Straith of Winnipeg who gave it to B.L. St. John. Aircraft slowly disintegrated at Kirkfield Park Aerodrome, Winnipeg. Fabric pieces remain only.

6149/18. From Dieppe via SS *Venusia* May 23, 1919.

7611/18. From Dieppe via SS *Venusia* May 23, 1919.

Phalz D.XII

2855/18. Via SS *War Peridat* July 7, 1919. To Acadia University, Wolfville, N.S., from Camp Borden May 14, 1920. Destroyed by fire December 1920.

Rumpler

9928/18. Type unspecified but likely a C.VII Rubilt. Via SS *War Peridat* July 7, 1919. To Mayor of Winnipeg, from Camp Borden, August 12, 1920. Destroyed in windstorm in fall of 1920 at St. Charles Aerodrome, Winnipeg.

More O.A.W.-built Fokker D.VIIs were assembled and flown at Leaside in August 1919 than noted above. The O.A.W.-built D.VIIs did not have their serials painted on them and while these aircraft can be seen in photographs they cannot be specifically identified.

It seems likely the above list may be incomplete for the following reason. The foundation of the list is the shipping record of the Canadian War Trophies Board with only the occasional other record being found. Even so, three specimens, Fokker D.VIIs 6810/18, 6842/18 and 8493/18, are recorded as being in Canada without a shipping record. Two Fokkers, likely D.VIIs, of unrecorded serials were listed as being shipped on the SS *War Peridat* on July 7, 1919, and are likely two of the three machines noted above. The wings of Fokker 693 (sic) were noted as being shipped from Dieppe on May 23, 1919, on the SS *Venusia* and this is believed to be in error for another machine shipped on that date. This indicates that at least one machine was brought to Canada without a shipping record. That being the case it seems possible that there might be others too.

The Snipe story began in the summer of 1917 when the Sopwith Aviation Co. started the design of a successor to its popular Camel. The first prototype appeared in the late summer and was a single-bay biplane with the 150 hp Bentley B.R.1 engine. In November a second prototype appeared with the 230 hp Bentley B.R.2 engine and increased dihedral; it was otherwise similar to the first, and about three machines were made to this design. These were later fitted with two-bay wings and one went to France in March 1918 for evaluation by pilots of Nos. 43 and 65 Squadrons.

The type was ordered into production and Sopwith and six other firms received orders for a total of 1,700 machines. It was intended that many of these would be produced as a more powerful version to be known as the Sopwith Dragon, which consisted of the Snipe airframe fitted with the 360 hp A.B.C. Dragonfly engine. Unfortunately the Dragonfly seldom could be persuaded to run satisfactorily so the type never entered squadron service, although over 200 Dragons were produced.

The Snipe first entered combat in late September 1918 with No.43 Squadron, and by October 31, 97 Snipes were in France with Nos. 43, 65 and 208 Squadrons, RAF, and No.4 Squadron of the Australian Flying Corps. Major W.G. Barker and his Snipe were only temporarily attached to No.201 Squadron, a Camel squadron, when his epic VC-winning combat took place.

After the Armistice many of the 1,700 Snipes on order were cancelled but a sizable number were completed. With the conservative, economizing thinking of the postwar years, the Snipe soldiered on as the RAF standard fighter, the great majority of Snipe squadrons serving abroad. Phasing out of the Snipe began in 1924 and the last left RAF service in 1927. Three Snipes came to Canada in the early postwar period. One was Barker's VC-winning

machine, another was entered in the New York-Toronto Race in 1919 but overturned just before the start, and the third served briefly with the Canadian Air Force at Camp Borden.

The Museum's Snipe, along with E6876 and another, was bought in 1926 by Reginald Leigh Dugmore Denny and brought to California. In about 1926 a fourth Snipe was brought to the United States, apparently by Arthur Le Baron. Another Snipe (or Dragon) was at McCook Field for evaluation by the US Army Air Service. Five Snipes came on the British civil register, twelve served with the Brazilian Army Aviation and at least one captured Snipe served with the Russian air service. These, together with the three Canadian machines, appear to be all the Snipes that were used in other than RAF service.

Reginald Denny, the importer of the three Snipes to the United States, was born in England in 1891. His father was an actor and "Reggie" followed him onto the stage in 1899 at the age of eight. He emigrated to the United States in 1911 and shortly made theatrical tours of India and the Orient before returning to England in 1914. He served in the British Army during 1917 and transferred to the RAF in 1918. He is reported as both an observer and pilot in the RAF before being discharged in 1919.

Denny then returned to the United States, settled in California and resumed his acting career. Over the years he appeared in productions on stage, screen and, later, television. However, he retained his interest in aviation and became interested in model aircraft. He sold model aircraft and developed radio-controlled models which were used as military targets. These were made by the Radio Plane Co. from 1940 to 1948 with Denny as Vice President and Secretary.

He was also a member of the 13 Black Cats organiza-

One of Denny's three Snipes forms the background for five of Hell's Angels *stunt pilots and its two male stars. Left to right: Ralph Douglas; Leo Nomis; Franke Clarke, Chief Pilot; the two stars, James Hall and Ben Lyons; Frank Tomick; Roy Wilson.* (NAM 11893)

tion shortly after its formation. It was formed after the well-known stunt pilot Richard V. Grace failed to show to perform an exhibition at the opening of Burdette Field in Los Angeles in late 1924. "Bon" MacDougall, Kenneth "Fronty" Nichols and "Spider" Mattlock stepped in to fill the gap and MacDougall flew a series of stunts while the others walked the wings. It was then thought that a market for stunt flying existed, both for air shows and for motion pictures, so the 13 Black Cats organization was formed to fill the need. They published a standard list of stunts with fees ranging from a maximum of $1,500 for blowing up an aircraft in mid-air and parachuting out to a single parachute jump for $80 and spinning an aircraft down, apparently on fire, but without crashing for $50.

Members of the 13 Black Cats are reported as Howard Batt, Odie Carter, Reginald Denny, Arthur E. Goebel, Sam Greenwall, Albert Johnson, "Wild Bill" Lind, "Spider" Mattlock, Herd McCelland, "Bon" MacDougall, Kenneth "Fronty" Nichols, Fred Osborne, J. Phillips, Paul E. Richter, Morrison Strapp and Ivan Unger. Denny was listed as an honorary member in deference to his studio, which forbade him to fly, but it is written that "on more than one occasion [he] evaded the studio watchdogs and joined in the escapades."

Howard Batt probably became the best known of the group as a stunt pilot, and he and Al Johnson flew in the film *Hell's Angels*, with Johnson losing his life during the filming. Batt later entered the general aviation business in Los Angeles. Arthur Goebel is best known as the winner of the ill-fated Oakland-Honolulu Dole Race in 1928 but was also a well known military pilot. Paul Richter became Executive Vice President of Transcontinental and Western Air (later Trans World Airlines) via Standard Airlines and Western Air Express. The organization disbanded in 1929.

It was almost certainly through his Black Cat association that Denny got the idea of importing the three Snipes. At the time civil aircraft were not registered in the United States so it was not until 1928 that they were assigned identity numbers 6636, 6637 and 6638, with the latter number assigned to E6938, now in the Museum. What became of the other two Snipes is not known but various accounts have been printed. One account states that one was burned and the other destroyed in an unknown manner. Another account says both were burned. Another states that both were given to the Los Angeles Educational Department and used in schools. Another account incorrectly states that one became E8100* now owned by Cole Palen of Rhinebeck, New York. That Snipe was apparently imported by Arthur Le Baron and sold to Leo Langevin, Binghamton, New York, and registered to him as 6949 in July 1928. Then it was sold to Myron A. Romberger, Endicott, New York, in September 1930, who in turn sold it in August 1932 to Roosevelt Field for their Museum, from which Palen bought it in the 1950s.

All three of Denny's Snipes appeared in the film *Hell's Angels*, but only in ground scenes† which are often omitted in some versions of the film. Why they were not flown is uncertain. One *Hell's Angels* pilot, Frank Tomick, stated that the Snipes were "too hot" for the other aircraft in the film. However, their specifications do not seem to indicate this. Possibly, their two-bay wings did not seem to fit with the other single-bay fighters used in the film. *Hell's Angels* was, of course, the most elaborate of all World War I motion pictures, with 78 pilots, an aviation ground crew of 150 and 40 aircraft. The well-known stunt pilot Frank Clarke was chief pilot in charge of flying operations. It is frequently said that Denny's Snipe (or Snipes) appeared in other movies. This may be so, but as an avid viewer of World War I aviation films since the late 1920s the writer cannot verify this statement.

* Identification arbitrarily assigned by Palen.

† For a rare still photograph of the three Denny Snipes in *Hell's Angels* see *Cross & Cockade*, Vol.6, No.3, p.25, illustration 21.

The Sopwith Snipe, still with its original RAF markings of the 1920s, on display March 11, 1950, at the National Orange Show at San Bernadino, California. It was this display that aroused Jack Canary's original interest in the Snipe. (KMM/A.U. SCHMIDT)

The Museum's Snipe, E6938, was built by Nieuport & General Aircraft Ltd., Cricklewood, England, the second of a batch, E6937-E7036, built in early 1919. Its service use is unknown but its aluminum finish indicates postwar use, likely abroad with most other RAF Snipes of the time. However, Jack Canary has written that it was built by Ruston, Proctor and Co. at Lincoln, accepted on September 22, 1918, and assigned to Farnborough for armament tests. These statements have been frequently repeated but cannot be substantiated.

Sometime, probably about the mid-1930s, Denny placed E6938 on loan to the Los Angeles County Museum, where it formed part of the science exhibit section in the basement. About 1942 it is said the L.A. Museum lost interest in the science section and the Snipe was sent to the USAAC's Norton Field under the care of the Public Information Officer, where it remained until after the end of World War II. Part of the time it was crated and part of the time it was assembled. In March 1950 it was assembled and displayed at the National Orange Show at San Bernardino, California, along with several other aircraft.

Apparently it was the sight of the Snipe at the show badly needing tender loving care that inspired Jack Canary to do something about it. First he tried the Los Angeles County Museum with no results, then the USAF with no results. In early 1953 there was an air show at the

Los Angeles International Airport and the air force agreed to have the Snipe exhibited there. Just what happened next is not entirely clear. J.J. Sloan wrote that Canary actually got permission to take the aircraft before it went on display, but George Holmes, who is believed to have the correct version, wrote that after the air show "the Snipe was moved a few blocks – and put behind a fence (thank God) – behind a trade school." This was beside Sepulveda Boulevard that Jack used to travel each day on his way to work. For some time he watched the Snipe deteriorating in the open.

Finally, Jack reached the Museum authorities and attempted to buy the aircraft with no success. Then it was agreed that he would restore the aircraft at no cost to the Museum and that the Air Force Association would provide all material and some labour. However, no assistance was received, and after some time Jack's lawyer threatened to place a lien on the machine. As a result the Museum agreed that Jack could have the aircraft as apparently they had little interest in it. At the time it was not realized that the Museum did not have clear title to the Snipe – it was only on loan to them from Denny.

Since Jack D. Canary, who was responsible for the fine restoration of the Snipe, and likely for its very survival, has now entered the story it is appropriate to give a brief biographical note about him. Jack was born in Milwaukee, Wisconsin, in 1916 and had his first flight in a "Jenny" in 1924, and aviation was his interest from then on. He got his pilot's licence in 1935 and operated a flying instruction business before World War II. He was a field service representative for North American Aviation during the war and served in China, where he had to bail out over enemy territory and was rescued by the USN. After the war he operated a fixed base aviation business at Phoenix, Arizona, doing aircraft modifications, cloud seeding and spraying. He then rejoined North American Aviation as a flight test engineer, later advance design engineer, and then went to Europe for the company. There he became involved with the production of the Bücker Jungmeister. At the same time he was acting as a consultant on the motion picture *Tora, Tora, Tora*. While ferrying a North American AT-6 for use in the film, Canary died in a flying accident on August 23, 1968, at Reading, Pennsylvania.

Having reached agreement with the Museum, Jack and a couple of friends disassembled the Snipe in September 1953 and moved it to his home in Pacific Palisades, near Santa Monica. During the initial period he took inventory of what items were missing from the

The Snipe's fuselage before restoration was started.
(NAM 11898)

The wings were rebuilt by the Northrop Aeronautical Institute at Inglewood, California.   (NAM 11892)

airframe and engine and began cleaning and repairing the parts on hand. At the same time he started the search for needed parts and information, which necessitated extensive correspondence with individuals and organizations in the United States and England.

Jack had numerous offers of help, some of which were helpful and others not. Some people took parts away to be cleaned and repaired and brought them back completed. Others did not bring them back, leaving Jack to round them up and complete the work. The wicker seat was taken away to be fixed or replaced and was never seen again. On the other hand D.W. "Dusty" Carter undertook to re-do the complete empennage at his home. This required replacing all the wood parts and returning the assemblies to Jack for covering. The Northrop Aeronautical Institute at Inglewood, California, rebuilt the wings, apparently as an instructional project, and it is believed all new wood was used. Warren Brodie and James J. Sloan both contributed many hours of work helping with the restoration and it is believed that other unidentified individuals also assisted.

Jack undertook the restoration of the fuselage and all its equipment, the undercarriage, the Bentley B.R.2 engine and the propeller. The propeller on the Snipe when received was an unknown American one, quite unsuitable in appearance and in performance, had anyone tried to use it. So Jack got a book on propeller design, studied it

and designed a new propeller for the B.R.2. He found a company, the Harmon-Porter Propeller Co., Fort Worth, Texas, who would make it to his design, his first and probably only propeller design. It was found to be entirely satisfactory on engine run-up and subsequent flight tests.

The engine was found, in Jack's words, to be in "overhaulable condition but lacking many parts and needed much rebuilding and repair work." In particular Jack wrote, "I had to completely rebuild the cam section as the engine at some time had swallowed a tappet which cracked both the intake and exhaust cam gear. It raised general hell with all keyways and spacers in the cam section and all this had to be rebuilt from scratch." Obviously returning the B.R.2 to serviceable condition was no sinecure but to aid Jack in this work was the fact that he had overhauled and repaired "a couple of dozen," conservatively estimated, different Gnome and Le Rhône rotary engines, undoubtedly more than any other person in recent years. The 80 hp Le Rhône in the Museum's Sopwith Pup and the 110 hp Le Rhône in the Museum's Nieuport both were overhauled by Jack and are both going strong, and the B.R.2 has performed flawlessly in the Museum's possession. Reginald Denny swung the propeller of the Bentley for the first run-up, with Jack in the cockpit, and this is said to have been recorded on film. The B.R.2 started on the first pull, as indeed it always has done since.

273

Jack Canary in his shop at Pacific Palisades, California. Behind him is the almost completed Snipe fuselage and its now airworthy Bentley B.R. 2 engine. Jack is adjusting the valve gear on an 80 hp Le Rhône rotary, one of many he restored to airworthy condition. (NAM 12305)

Publicity given to Jack's restoration project resulted in both missing magnetos, the cockpit nameplate and other missing parts being returned to him after having been "removed" while the Snipe was stored outside. Frank Courtney, the well-known British pilot, then in the United States, put Jack in touch with helpful people and organizations in England. Sir Sydney Camm of Hawker Aircraft Ltd. supplied the needed drawings. The PLM magnetos, originally made by the M-L Magneto Syndicate Ltd., were overhauled by their successors, Ignition and Electrical Services in England. The Dunlop Tire and Rubber Co. in England re-rimmed the wheels and installed new tires. S. Smith and Sons Ltd. produced instrument drawings and from these accurate and operable instrument reproductions were made, apparently by Smith. The compass was acquired in an exchange with the USAF Museum. The Flightex Co. donated all the needed fabric and North American Aviation helped generously, but unofficially; among other things they lofted all the fuselage formers. It will be seen that the restoration, although sparked and inspired by Jack, was aided in many indispensable ways by individuals and firms on both sides of the Atlantic. Without this assistance it is unlikely that the excellent outcome of the restoration could have been achieved.

When all structural work had been completed, it was likely Jack who recovered the aircraft, with assistance, but this is not recorded. Jack decided to refinish the Snipe as a machine of No.208 Squadron in 1919, then serving in Germany, rather than its original aluminum finish. After completion in the summer of 1960 it was widely acclaimed as the best restoration of a World War I aircraft ever carried out. Jack took it to Torrance Municipal Airport and flew it once from its grassy side. This flight and the satisfaction of seeing the Snipe so well restored after seven years of work were Jack's reward for his efforts.

The USAF at Wright-Patterson AFB was shortly having a reunion of World War I aviation personnel and arranged to borrow the Snipe for the occasion, and it was then loaned to the Museum for display. Jack had tied up a good deal of his funds in rebuilding the aircraft and, understandably, felt he had to retrieve them. An agreement was reached with Denny. The Snipe was offered for sale and bought by the Canadian War Museum, as related elsewhere in this book.

Late in 1987 the aircraft was re-covered and minor changes were made to ensure that it was in its best form for the opening of the Museum's new building in June 1988.

REFERENCES

— "Jack D. Canary Obituary." In *AAHS Newsletter*, No.8
— "Field Trip to Photograph the 'Snipe.'" In *Aeronautica*, Vol.2, No.2, April-June, 1950.
— Barclay, John Allen. "Hollywood's 13 Black Cats." In *AAHS Journal*, Vol.28, No.3/4, p.243.
— Bruce, J.M. "Sopwith Snipe – The RAF's First Fighter." In *Air Enthusiast International*, April and June, 1976.
— Canary, Jack D. Letter to Frank Straad, April 7, 1959.
— Canary, Jack D. "New Feathers for a Snipe." In *Aeronautica*, Vol.8, No.1, Spring 1958.
— Carter, D.W. Letter to the author, October 17, 1983.
— Fisk, Gary. "Summary of Antique and Replica Aircraft in California." In *AAHS Journal*, Vol.5, No.2, Spring 1960.
— Hatfield, D.D. In *Los Angeles Aeronautics*, Northrop University.
— Holmes, George B. Letter to the author, June 21, 1979.
— Levy, Howard. "Cross and Cockade Fly Again." In *Air BP*, No.23.
— Nichols, Ken ("Fronty"). "The Black 13." In *Argosy Magazine*, April 1958.
— Rigdon, W. *Biographical Encyclopedia & Who's Who of American Theatre*.

List of Aircraft Specimens in the National Aviation Museum

The following list records not only complete specimens but also some major components held. While most of the information recorded will be readily interpreted, the following notes are provided to clarify the terms and the codes used. The entire listing is up to date as of the opening of the new Museum building in 1988.

Design Firm (Maker)

This records the designing company and, where the maker was other than the design firm, the maker's name. Other firms were, and are, frequently given licences to produce specific types of aircraft, especially in wartime, and often more examples of a type have been made by licensees than by the original design firm.

Type and Name

Most aircraft types are given a number by their design firm and sometimes by their military or naval users as well. Sometimes types with no official name have been given an unofficial one by their crews, often with a good deal of affection. Unofficial names are given in quotation marks. Reproduction aircraft are indicated by (R) in the type column.

Status/Location

This column defines the state and location of each specimen upon the opening of the new building in 1988. The coding used is as follows:

A	Flown on occasion and may be away on a flight demonstration at those times
C	Under restoration
L	On loan to another museum
PV	On view to the public
S	Stored
X	A suffix indicating that the specimen is in colours and/or markings other than those it bore during its active life
/R	At Rockcliffe
/T	At Toronto
/W	At Canadian War Museum, Ottawa
/WC	At Western Canada Aviation Museum, Winnipeg

Construction Number

A number assigned by the maker to the specimen remains unchanged regardless of the number of times the specimen changes owners or in what country it is registered. Some firms consider the military serial assigned as their construction number.

Military Serial and Civil Registration

These identifying marks serve to distinguish the specimen from others within the service or country. As a specimen may serve in both military and civil use during its life as well as in different countries, it may bear several different identities over the years: e.g., the Museum's D.H.80A Puss Moth has been 8877 of the USN, HM534 of the RAF, G-AHLO on the British civil register, and CF-PEI on the Canadian civil register. However, where no parentheses are found in one of the two columns, it indicates the specimen is, or has been, used for flight demonstrations by the Museum; the Canadian civil registration is displayed to satisfy flying regulations.

Year Manufactured

This gives the last two figures of the year in which the specimen was made. Where an exact date has not been established the date is given as c.XX.

Design Firm (Maker)	Type	Name	Status/ Location	Construction Number	Military Serial	Civil Registration	Year Mfd.
A.E.A. (RCAF)	(R)	*Silver Dart*	PV/R	None	None	None	57
A.E.A. (RCAF)	(R)	*Silver Dart*	S/R	None	None	None	57
A.E.G.	G.IV		PV/R		574/18		18
Aeronca	C-2		SX/R	9	(N525V)	CF-AOR	29
Airspeed	A.S.65	Consul	S/R	4338	(PK286)	G-AIKR	44
Auster	A.O.P.6		S/R	TAV22IV	VF582	(CF-KBV)	48
Avro	504K		SXA/R		G-CYCK	CF-CYC	c.17
Avro (RCAF)	504K(R)		PV/WC	None	G-CYEI		67
Avro	504K		PVX/R		(G-CY)FG	(H2453)	c.18
Avro	616	Avian IVM	S/R	314	(134)	CF-CDQ	30
Avro (MacDonald Bros.)	652A	Anson V	PV/R		12518		44
Avro (Victory Aircraft)	683	Lancaster X (nose)	PV/R		Unknown		c.44
Avro (Victory Aircraft)	683	Lancaster X	PV/R		KB944		44
Avro Canada	CF-100	Canuck Mk.5	S/R		100785		58
Avro Canada	CF-100	Canuck Mk.5	S/R		100757		58
Avro Canada	C-102	Jetliner (nose)	PV/R			CF-EJD-X	49
Avro Canada	CF-105	Arrow 2 (nose)	PV/R		25206		59
Bell	HTL-6 (47G)	Sioux	PV/R	1367	1367		55
Bellanca	CH-300	Pacemaker	PV/R	181	(NC196N)	CF-ATN	30
Blériot (Calif. Aero)	XI		S/R	None	None	None	11
Boeing	247D		PV/R	1699	(N7638)	CF-JRQ	33
Bristol (Fairchild Canada)	149	Bolingbroke IVT	SX/R		9892		42
Bristol	156	Beaufighter X.T.F.	S/R		RD867		43
Canadair	CL-28	Argus 2	S/R		10742		60
Canadair	CL-84-1	Dynavert	S/R	3	8402		71
Canadian Vickers		Vedette V & Va (two hull portions only)					—
Cessna	T-50	Crane I	S/R	2226	8676		42
Consolidated (Can. Vickers)	PBY-5A	Canso A 2SR	S/R	CV423	11087		44
Consolidated (Ford)	B-24L	Liberator G.R. VIII	SX/R	(44-50154)	11130	(HE773)	44
Curtiss	HS-2L		PV/R		(A1876)	G-CAAC	18
Curtiss	JN-4 (Can.)	"Canuck"	PVX/R		C227	(39158)	18
Curtiss	18	Seagull	PV/R	Unknown	None	None	c.22
Curtiss	P-40D	Kittyhawk I	S/R	18780	1076	(AL135)	42
Czerwinski/Shenstone		Harbinger	S/R	C1		C-FZCS	75
De Havilland	D.H.60X	Moth	PV/R	630		G-CAUA	28
De Havilland	D.H.80A	Puss Moth	PV/R	2187	(HM534)	CF-PEI	31
De Havilland (DHC)	D.H.82C2	Menasco Moth I	PV/R	1052	4861		41
De Havilland (DHC)	D.H.82C2	Tiger Moth I	S/R	724	(4394)	CF-FGL	41

Design Firm (Maker)	Type	Name	Status/ Location	Construction Number	Military Serial	Civil Registration	Year Mfd.
De Havilland (DHC)	D.H.98	Mosquito XXB	S/R		KB336		44
De Havilland	D.H.100	Vampire I	S/R		TG72		46
De Havilland	D.H.100	Vampire 3	S/R		17074		47
De Havilland Canada	DHC-1B2	Chipmunk 2	S/R	208-246	12070	CF-CIA	56
De Havilland Canada	DHC-2	Beaver 1	PV/R	1		CF-FHB	47
De Havilland Canada	DHC-3	Otter	S/R	370	9408		60
De Havilland Canada	DHC-6	Twin Otter Series 100	S/R	1	CF-DHC-X		65
Douglas	DC-3		PV/R	6261	(43-1985)	C-FTDJ	43
Douglas	C-47B	Dakota IV	S/R	32922	KN451	(44-76590)	45
Douglas Canadair	C-54GM	North Star 1 ST	S/R	122	17515		48
Fairchild	FC-2W-2		PVX/R	128	(NC6621)	G-CART	28
Fairchild (Canada)	82A		S/R	61		CF-AXL	37
Fairchild (Fleet Canada)	PT-26B	Cornell III	S/R	FC239	10738		43
Fairey		Battle IT	S/R		R7384		39
Fairey		Swordfish I.T	PVX/R	Unknown	Unknown		c.43
Farman, Maurice (Airco)	S.11	"Shorthorn"	PV/R			VH-UBC	c.16
Fleet (Fleet Canada)	2/7	(parts only)	S/R	Unknown		Unknown	c.30
Fleet (Fleet Canada)	16B	Finch II	PV/R	408	4510	(N1327V)	40
Fleet Canada	50K	Freighter	S/R	202	(800)	CF-BXP	39
Fleet Canada	80	Canuck	S/R	149		CF-EBE	46
Fokker	D.VII		C/R	3659	10347/18	(N1178)	18
Found	FBA-2C		S/R	4		CF-OZV	62
Hawker		Hind I	S/R		L7180		37
Hawker (Canadian Car)		Hurricane XII	PV/R		5584		42
Hawker		Sea Fury F.B.11	PV/R		TG119		48
Heinkel	He162A-1	Volksjäger	SX/R		120076		45
Heinkel	He162A-1	Volksjäger	S/R		120086		45
Hispano (Bf 109G2)	HA-1112-M1L	Bouchón	PV/R	164	C.4K-114		c.55
Junkers	J.I(J4)		S/R		586/18		18
Junkers	W.34f/fi		PV/R	2718		CF-ATF	32
Lockheed	10A	Electra	PV/R	1112	(1526)	CF-TCA	37
Lockheed	12A	Electra Jr.	S/R	1219		CF-CCT	37
Lockheed (Canadair)	T-33AN	Silver Star 3	S/R	T33-574	21574		57
Lockheed	F-104A	Starfighter	PV/R	183-1058	12700	(56-770)	56
Lockheed	1329	Jetstar 6	S/R	5018	(N9190R)	C-FDTX	62
McDonnell	F2H-3	Banshee	PV/R	174	126464		53
McDonnell	CF-101B	Voodoo	S/R	518	101025	(57-340)	c.57
McDowall		Monoplane	PV/R	None	None	None	15

Design Firm (Maker)	Type	Name	Status/ Location	Construction Number	Military Serial	Civil Registration	Year Mfd.
Messerschmitt	Me 163B-1a	Komet	PV/R		191095		45
Messerschmitt	Me 163B-1a	Komet	S/R		191916		45
Nieuport	12		SX/R		A4737		c.15
Nieuport (Swanson)	17(R)		S/R	None	B1566	CF-DDK	63
Noorduyn	UC-64A	Norseman VI	PV/R	136	787		43
North American	AT-6	Harvard II	PV/R	66-2565	2532		40
North American	AT-6	Harvard II	S/R	81-4107	3840		41
North American (Can. Car)	T-6J	Harvard 4	S/R	CCF178	20387	CF-GBV	52
North American	TB-25L	Mitchell 3PT	S/R	108-47453	5244	(44-86699)	44
North American	P-51D	Mustang IV	PVX/R	122-39806	9298	(44-73347)	44
North American (Canadair)	F-86E	Sabre 6	PV/R	1245	23455		55
North American (Canadair)	F-86E	Sabre 6	S/R	1441	23651		56
Northrop (Can. Vickers)		Delta II (parts only)	S/R	183	673		37
Piasecki	HUP-3		PV/R		51-116623		c.51
Pitcairn-Cierva	PCA-2	Autogiro	S/R	B-18	(NR26)	NC2624	30
R.A.F. (Brit. & Colonial)	B.E.2c		S/R		4112		15
Sikorsky	R-4B		S/R		43-46565		43
Sikorsky	(S-51) H-5	Dragonfly	S/R		9601		47
Sikorsky	(S-55) HO4S-3		PV/R		55877		c.53
Sopwith (Neal)	(R)	Pup	S/R	C552	B2167	CF-RFC	67
Sopwith (Swanson)	(R)	Triplane	PV/R		N5492	CF-CBM	66
Sopwith (Hooper)	2F.1	Camel	PVX/W		N8156		18
Sopwith (Nieuport & Gen)	7F.1	Snipe	PVX/R		E6938		19
Sopwith	7F.1	Snipe (fuselage)	PV/W		E8102		18
Spad (Mann & Egerton)	7		PV/R		B9913		c.17
Stearman	4EM	Junior Speedmail	PVX/R	4021	(NC784H)	CF-AMB	30
Stinson	SR	Reliant	PV/R	8717	(NC13464)	C-FHAW	33
Stits (Hopkinson)	SA-3A	Playboy	S/R	5501	(CF-IGK-X)	C-FRAD	55
Supermarine (Castle Brom.)		Spitfire IIB	PV/W	CBAF711	P8332		40
Supermarine (Castle Brom.)		Spitfire IX L.F.	PV/R	1X2161	NH188	(CF-NUS)	44
Supermarine (Castle Brom.)		Spitfire XVIe	S/R	1X4424	TE214	(TE353)	45
Taylor	E-2	Cub	PV/R	289	(NC15399)	C-GCGE	35
Travel Air	2000		S/R	720	(C6281)	CF-AFG	29
Vickers	757	Viscount	S/R	270		CF-THI	57
Westland		Lysander III	PVX/R		R3003		—
Wills Wing	XC-185	(Hang glider)	S/R	None	None	None	77
Zenair	CH300	Tri-Zenith	S/R	300		C-GOVK	78

278

List of Aircraft Powerplants
in the National Aviation Museum

This list includes engines installed in aircraft specimens and those held as separate artifacts. In some cases multiple specimens of a type are held, allowing an engine to be exhibited separately near an aircraft that has the same engine installed and therefore usually concealed.

Reciprocating engines

A.D.C. Cirrus II
Aeronca E-113C
Alfa-Romeo 115-I
Allison V-1710
Anzani (fan type) 25 hp
Anzani 6 cyl. radial 40/45 hp
Armstrong-Siddeley Lynx IV
Armstrong-Siddeley Genet II
Armstrong-Siddeley Genet
    Major 5 cyl.
Armstrong-Siddeley Jaguar VIB
Armstrong-Siddeley Cheetah IX
Beardmore 160 hp
Bentley A.R.I./B.R.I.
Bentley B.R.2
Benz BZ IIIa 185 hp
Benz BZ IV 200 hp
B.M.W. Type 1321
Bristol Jupiter IV
Bristol Jupiter X
Bristol Mercury VIII
Bristol Mercury XV
Bristol Mercury XX
Bristol Perseus X
Bristol Pegasus XVIII
Bristol Pegasus XXX

Bristol Centaurus XVIIIC
Bristol Hercules XVII
Clerget 9Z 110 hp
Clerget 9B 130 hp
Continental A-40
Continental A-65
Continental C-90
Continental R-975-46
Curtiss *Silver Dart*
Curtiss L
Curtiss O
Curtiss OX-5
Curtiss OXX-6
Curtiss C-6A
Curtiss V-1570 Conqueror
Daimler Mercedes D.IIIav 160 hp
Daimler Mercedes D.IIIavu 180 hp
Daimler Mercedes D.IVb 260 hp
Daimler-Benz DB 600
D.H. Gipsy III
D.H. Gipsy Major
D.H. Gipsy Major 1C
D.H. Gipsy Major 7
D.H. Gipsy Major 10
Duryea 16 hp
Dutheil & Chalmers 20 hp

Elbridge Aero Special 60 hp
Fairchild Caminez 447C
Fiat RT12
Franklin 4AC-176B2
Franklin 6V4-200-C32
Gibson 60 hp
Gnome Omega 50 hp
Gnome Delta 100 hp
Gnome N 160 hp
Gnome-Rhône 1417
Hall-Scott A-7A
Hall-Scott L-6
Hispano-Suiza 180 hp
Jacobs L-4MB
Jacobs L-5MB
Jacobs L-6MB
Jacobs R-915-7
Junkers L.5
Junkers Jumo 211
Kinner B-5
Kinner K-5
Kinner R/2
Kirkham B-6
Le Rhône 9C 80 hp
Le Rhône 9Jb 80 hp
Le Rhône g Jb 110 hp

Liberty 12
Lycoming R-680-9
Lycoming R-680-17
Lycoming 0-235-C
Lycoming IGS 0-480-A1E6
Lycoming 0-540-A1D
Lycoming IGO-540
Menasco D.4 Super Pirate
Napier Lion II
Napier Sabre
Piaggio P.VII.C
Pratt & Whitney Wasp C
Pratt & Whitney Wasp SC
Pratt & Whitney
    R-985 AN 14B Wasp Jr.
Pratt & Whitney R-1340-47 Wasp
Pratt & Whitney R-1340-AN-1 Wasp
Pratt & Whitney S1H1-G Wasp
Pratt & Whitney
    SB-4-G Twin Wasp Jr.
Pratt & Whitney R-1830 Twin Wasp
Pratt & Whitney
    R-2800-31 Double Wasp
Pratt & Whitney R-4360 Wasp Major
R.A.F. 1A
Ranger 6-440-C
Renault Vee-8 80 hp
Rolls-Royce Hawk II
Rolls-Royce Eagle VII
Rolls-Royce Eagle IX
Rolls-Royce Kestrel V
Rolls-Royce Merlin II
Rolls-Royce Merlin III
Rolls-Royce Merlin IV
Rolls-Royce Merlin XX
Rolls-Royce (Packard) Merlin 29
Rolls-Royce (Packard) Merlin 33
Rolls-Royce Merlin 35

Rolls-Royce (Packard) Merlin 68A
Rolls-Royce (Packard) Merlin 69
Rolls-Royce Merlin 76
Rolls-Royce Merlin 114A
Rolls-Royce (Packard) Merlin 224
Rolls-Royce (Packard) Merlin 226
Rolls-Royce Merlin 500
Rolls-Royce Merlin 620
Rolls-Royce Griffon
Salmson 9AB
Siddeley-Deasy Puma
Warner Super Scarab
Wolseley Viper II 210 hp
Wright J-4B Whirlwind
Wright J-5A Whirlwind
Wright J-6-9 Whirlwind
Wright R-975-11 Whirlwind
Wright R-1300 Cyclone 7
Wright GR-1820-G202A Cyclone
Wright R-1820 F52 Cyclone
Wright R-2600-29A Cyclone 14
Wright R-3350 Cyclone 18

Turbojet Engines

Avro Canada Chinook
Avro Canada Orenda 2
Avro Canada Orenda 11
Avro Canada Orenda 14
B.M.W. 109-003E-1
D.H. Goblin 2
D.H. Ghost 1
General Electric J47-GE-13
General Electric J79-OEL-7
Hispano-Suiza 104 (Nene)
Junkers Jumo 004
Metrovick F.2 Berl
Orenda 1R

Orenda Iroquois
Pratt & Whitney JT3D
Pratt & Whitney JT3P
Pratt & Whitney JT12-6
Pratt & Whitney J57-F55
Rolls-Royce Derwent 1
Rolls-Royce Derwent 5
Rolls-Royce Derwent 8
Rolls-Royce Derwent RB9
Rolls-Royce Derwent RB37-9
Rolls-Royce Derwent 502
Rolls-Royce Nene 1
Rolls-Royce Nene 3
Rolls-Royce Nene 10
Rolls-Royce Avon XC-100
Rolls-Royce Avon 1
Rolls-Royce Avon RA2
Rolls-Royce Conway RCo 12
Westinghouse J34-WE-34

Turboprop Engines

Lycoming T.53
Pratt & Whitney Canada PT-6-120
Rolls-Royce Dart 505
Rolls-Royce Dart 506

Pulsejet Engines

Argus As 014

Ramjet Engines

Marquardt RJ43-MA-7

Rocket Engines

Aerojet YLR91-AJ3AJ
Thiokol XM-51
Walter 109.509C

List of Acronyms

This list of acronyms and abbreviations includes all the current Canadian Forces rank abbreviations. Current rank abbreviations do not take periods (as in Col Smith) while the older style used periods (as in Col. Smith).

A/C	Air Commodore	CWO	Chief Warrant Officer
A/C/M	Air Chief Marshal	DFC	Distinguished Flying Cross
AC1	Aircraftman, 1st Class	DHC	De Havilland Canada
AC2	Aircraftman, 2nd Class	DND	Department of National Defence
A.D.C.	Aircraft Disposal Co. (Airdisco)	DOT	Department of Transport
A.E.A.	Aerial Experiment Association	EFTS	Elementary Flying Training School
A.E.G.	Allgemeine Elektrizitäts Gesellschaft	F/L	Flight Lieutenant
AFB	Air Force Base	Flt Sgt	Flight Sergeant
AITA	Air Industries and Transport Association	F/O	Flying Officer
A/M	Air Marshal	FS	Flight Sergeant
AOP	Air Observation Post	F/S/L	Flight Sub-lieutenant
A/V/M	Air Vice-Marshal	G/C	Group Captain
BCATP	British Commonwealth Air Training Plan	Gen	General
BGen	Brigadier-General	HMAS	Her (His) Majesty's Australian Ship
CAF (1919-23)	Canadian Air Force	HMCS	Her (His) Majesty's Canadian Ship
CAF (1968 on)	Canadian Armed Forces	HMS	Her (His) Majesty's Ship
CAI	Canadian Aeronautical Institute	IAF	Indian Air Force
Capt	Captain	IAS	Institute of Aeronautical Sciences
C de G	Croix de Guerre	LAC	Leading Aircraftman
CF	Canadian Forces	LGen	Lieutenant-General
CPA	Canadian Pacific Airlines	Lt	Lieutenant
Col	Colonel	Lt. Cdr.	Lieutenant-Commander
Cpl	Corporal	Lt. Col.	Lieutenant-Colonel
CWM	Canadian War Museum	Maj	Major

Maj.-Gen.	Major-General	R.A.F.	Royal Aircraft Factory
MC	Military Cross	RCAF	Royal Canadian Air Force
MCpl	Master Corporal	RD	Repair Depot (RCAF)
MGen	Major-General	RNAS	Royal Naval Air Service
MOT	Ministry of Transport	RNZAF	Royal New Zealand Air Force
MWO	Master Warrant Officer	R.O.M.	Royal Ontario Museum
NAC	National Aeronautical Collection	SFTS	Service Flying Training School
NACA	National Advisory Committee for Aeronautics	Sgt	Sergeant
		S/L	Squadron Leader
NACPAC	National Aeronautical Collection Policy Advisory Committee	Spad	Société anonyme pour l'Aviation
		SS	Steamship
NAM	National Aviation Museum	STOL	Short Take-off and Landing
NASM	National Air and Space Museum	TAM	The Aeronautical Museum (at NRC)
NATO	North Atlantic Treaty Organization	TCA	Trans-Canada Air Lines
NMC	National Museums of Canada	USAAC (1926-41)	United States Army Air Corps
NMST	National Museum of Science and Technology	USAAF (1941-47)	United States Army Air Force
NRC	National Research Council	USAF (1947 on)	United States Air Force
O.A.W.	Ostdeutsche Albatros Werke	USAS (1918-20)	United States Air Service
OCdt	Officer Cadet	USMC	United States Marine Corps
OPAS	Ontario Provincial Air Service	USN	United States Navy
OTU	Operational Training Unit	USS	United States Ship
P&W	Pratt & Whitney	VC	Victoria Cross
PAC	Public Archives Canada	VTOL	Vertical Takeoff and Landing
P/O	Pilot Officer	WCAM	Western Canada Aviation Museum
Pte	Private	W/C	Wing Commander
PWA	Pacific Western Airlines	WO	Warrant Officer
RAAF	Royal Australian Air Force	WO1 (WO I)	Warrant Officer, Class I
RCN	Royal Canadian Navy	WO2 (WO II)	Warrant Officer, Class II
RFC	Royal Flying Corps	2 Lt	Second Lieutenant
RAF	Royal Air Force	2nd Lt.	Second Lieutenant

Index

EDITORIAL NOTE:
Page numbers in italics
refer to an illustration.

285

ON THE MAKING OF THIS BOOK

Project Manager:	**Wendy McPeake**
English Editors:	**Robin Brass**
	Wendy McPeake
	Hladini Wilson
French Editor:	**Adèle Lessard**
Designer:	**Frank Newfeld**
Typesetter:	**Fleet Typographers Limited**
Printer:	**D.W. Friesen Printers**